EDAPHOS

Dynamics of a Natural Soil System

by

Paul D. Sachs

Newbury, Vermont

EDAPHOS
Dynamics of a Natural Soil System
by Paul D. Sachs

THE EDAPHIC PRESS
P.O. Box 107
Newbury, Vermont 05051

Library of Congress Catalog Card Number 93-70568

ISBN 0-9636053-0-5

ABOUT THE AUTHOR

In 1983, when Paul founded North Country Organics*, he decided it was important to learn something about the soil if he was going to sell natural fertilizers. However, as time went on, the objective changed from an obligatory curiosity to a quest of passion. As the information accumulated, Paul began to see metaphoric similarities between the soil system and all other terrestrial dynamics including the social and economic structures established by human civilization. He decided that many analogies from the soil such as competition, growth, adaptation and symbiosis could be applied to all of life's relationships, including running a successful business.

After only two years of study, Paul became known as a source of good information for people involved in organic crop production or land care. Requests for articles and seminar appearances began to increase. As time passed, his knowledge about natural soil systems began to encumber his work (i.e. selling natural land care supplies) because so much time was being spent on consultation. In 1991, Paul began work on **EDAPHOS** to help diffuse the tremendous amount of valuable information out to the people who need it the most. This work represents almost ten years of his research on the subject of soil system dynamics.

* North Country Organics is a Bradford, VT based supplier of natural fertilizers and soil amendments for commercial use.

DEDICATED TO MARIAN
*for teaching me that
my limitations are almost always
self inflicted.*

ACKNOWLEDGMENTS

Special thanks to Robert Parnes, Dr. Leonard Perry, Grace Gershuney, Peter Luff, Lewis Hill and Karen Idoine for their constructive comments and suggestions on how to improve EDAPHOS. Thanks also to Penelope Hoblin for her help on the cover and her willingness to give detailed advice on a moment's notice.

Extra special thanks to Ruth, Dave and Jeff for their love and support and for putting up with my alarm clock going off every morning at 5:00 am for almost two years.

EDITOR'S NOTE

As I read through Paul's manuscript for the first time, I was struck by the sincerity of his voice and by his genuine concern for our world. After my second reading however, I was struck more forcefully by the relevance of the information he was giving to my own life and the world around me.

Edaphos helped me to realize, not only the need for a life-style and attitude change toward the physical world, but it also helped me to understand why this need exists and how to go about making some of these changes.

Paul's simplification of chemistry and language helps the layperson to understand a subject that is, all too often, kept out of reach from him. The analogies that are drawn between human systems and relationships and those of the soil bring the soil to life... a characteristic which can not help but increase one's reverence for the earth we live with.

Sadly enough, *Edaphos* also helped me to realize that the soil, like the human, is mortal too. And if we do not take ownership for the practices we have concerning it, we will lose it.

It is refreshing to find that in a time when most want to take, take, take; and a time when we look for instant gratification and that quick fix, *Edaphos* offers a reasonable plan to give back some of what we have taken. Paul gives his reader valuable information not just specific to gardening, but to living life in general.

In *Edaphos*, Paul reinforces what the late Fred Franklin once advised my husband; *We should live each day of our lives as if it were the last and use our soil as if we were going to live forever.*

Wendy Goldsworthy

CONTENTS

INTRODUCTION   1

PART I - **IN PRINCIPLE**

CHAPTER 1
SOIL EVOLUTION   7

CHAPTER 2
**HUMUS: A STABLE
ORGANIC MATTER**   33

CHAPTER 3
WATER   53

CHAPTER 4
THE CONCEPT OF PESTS   65

PART II - **IN PRACTICE**

CHAPTER 5
**COMPOSTING & PRESERVING
ORGANIC MATTER**   85

CHAPTER 6
ORGANIC *VS* INORGANIC   107

CHAPTER 7
TESTING THE SOIL SYSTEM   139

CHAPTER 8
RELATIONSHIPS   169

GLOSSARY   173
INDEX   181

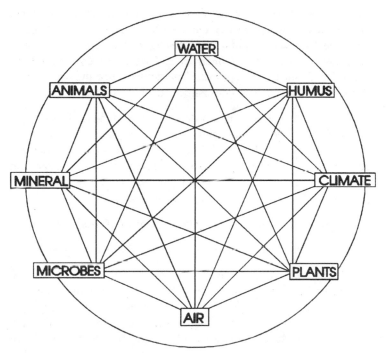

Inter-relationships of edaphic factors

INTRODUCTION

Edaphos is an ancient Greek word meaning ground or soil. The study of edaphology is technically the science of soil from the perspective of plant growth or plant production. This book is an analysis of the natural soil system that determines the environment for plants. However, plants, and all other living things, are also factors that influence their own environments.

Throughout this work the importance of each factor to the overall conditions of every environment is stressed as comparative but not emphasized as superlative. In an ecosystem formed by many different influences each becomes a critical component; if one factor is changed or eliminated the whole system can change.

Since life began changes in the environment have occurred. Many species have become extinct and many new ones have evolved. Judgment as to whether this natural evolution is good or bad cannot be made. However, human disturbance of various environments have caused numerous extinctions and have not encouraged the evolution of many new species. The depletion of the earth's biodiversity is the alarming consequence of this trend.

This book attempts to create a holistic sense of the earth as a single entity that functions as a result of all the physical and biological events that occur. Although life is incredibly diverse and intensely competitive, the biosphere seems to function like a large and complex living organism. All of its organic and inorganic components participate as essential parts of the whole.

The edaphic factors can be simplified into four basic categories:

The living • The dead • The minerals • The atmosphere

Within each category are countless variables that all have an immeasurable effect on the whole system. From these factors we can derive the theory: *Energy that affects anything will eventually affect everything else.* It is hard to imagine and impossible to predict how the future of a planet can be affected by subtle changes in the living and non-living components of its mass. However, eons of evolution prove that anything is possible.

Extraterrestrial factors such as earth wobble, solar flares, planetary alignment or universal contraction/expansion are not considered even though their significance is recognized.

The science of edaphology is actually the study of many different sciences including (but is not limited to) mineralogy, meteorology, biology/microbiology, botany, pedology, geology, chemistry, physics and hydrology. Mother nature began instruction in each of these fields a long time ago. However, in our specialized studies of various fields the relationships to other sciences are often ignored.

Our examination of the natural system is limited by the amount of information available to us. The tremendous volume of research that has accumulated over the years may only be a small fraction of what is yet to be learned from nature. Some of her secrets may never be discovered. Understanding the natural system must be tempered with humility i.e. the notion that there are portions of the system that we humans may never understand. An element of faith needs to be part of the land care formula. This is not to say that one should run into a forest naked to become *one-with-nature* but, instead, to respect the cause-and-effects of human intervention and work with the environmental limitations as much as possible.

In the human quest to isolate and concentrate nature's secrets, we have developed another edaphic factor that is changing the ecosystem of the planet. The human factor is very real and is causing changes in the environment that may or may not be harmful to the earth itself in the long term, but is clearly harmful to the existence of many different species of living organisms, including our own.

Edaphos is about the purpose and balance of all components of the environment whether animate or inanimate, organic or inorganic, pest or beneficial, essential nutrient or toxic heavy metal. If one believes that everything on earth has a purpose, then it follows that problems associated with agriculture and horticulture are due to imbalances, not to the existence of a specific factor that appears to be causing the problem.

At the beginning of the twentieth century scientists made incredible progress discovering the hidden functions and relationships of the soil's ecosystem. However, after World War II the world of agriculture and horticulture became convinced that crop production could transcend nature with magical nitrates left over from the production of bombs and bullets. The farmer learned how to make greater profits by eliminating rotations that would allow the soil to rebuild and restore the nutrients lost from crop removal. Unfortunately, they also found they needed to compliment their menu of synthetic nutrient with deadly biocides to combat any of nature's objections to their strategies. Eventually, it became another war. The purpose of this book is to rediscover the natural relationships that synergise the ecosystem and, hopefully, to help put an end to the war.

Our relationship with the soil can be analogous to our relationships with other people. Without knowing the intricate psychologies of the individuals we relate to, we can interact peacefully and productively, or choose to dominate the relationship with might and intimidation. The latter choice usually leads to a chain reaction of disputes and possibly violence where no one realizes a gain. The former choice, however, can be self sustaining for an indefinite period of time.

AUTHOR'S NOTES

This book is intended for anyone who has a desire to increase his understanding of elementary soil science. Terminology thought to be beyond conversational vocabulary is defined in the glossary located in the back of the book.

My use of the term *Mother nature* is with the utmost respect for both the earth and women. The historical roots of reffering to the planet in a maturnal manner is well documented and is normally expressing sincere homage.

Throughout the book, the masculine gender (i.e. his, him, he) is used when refering to humans. Historically, these pronouns have been used for generic expression in literature. No disrespect toward women is intended or implied. The use of *his/her or (s)he,* in the text is awkward and difficult to read.

Part I
IN PRINCIPLE

Chapter 1
SOIL EVOLUTION

A long, long time ago, the earth was a very different place. The atmosphere was filled with toxic gases and the climate was too hot for any living thing. The ground as we know it was barren and molten. The ultra-violet rays of the sun came beaming down with full force, unfiltered by ozone that hadn't yet formed.

As the planet began to cool, different types of rock formations appeared, followed by a weathering process that began to form our ancient soils (keep in mind, events that took millions of years to occur were just condensed into one sentence). Oceans formed and life began to appear in the form of microorganisms. These organisms brought about changes in the earth's atmosphere, such as a decrease in carbon dioxide and an increase in oxygen, that eventually made it possible for land life to evolve. This new atmosphere caused an evolution of new species of organisms, many of which could adapt to life outside of the ocean.

The relationship that formed between those prehistoric particles of weathered rock and the first organisms that inhabited dry land is the basis for today's soil and life as we know it.

WEATHERING

The weathering of rock mineral still occurs today as it has for millions of years. Physical forces such as rain, wind, frost, gravity, root activity and tectonic movement cause the systematic destruction of rock into smaller and smaller particles. Chemical reactions with organic acids, enzymes and water further pulverize what used to be mighty mountains into sand, silt and clay.

Water probably plays the biggest role in the weathering process. Rain pounding down on rock surfaces washes away loose particles that then become abrasives as they travel downward with the water's flow. Eventually more and more water joins together to find a common avenue, forming fast moving streams that cause even greater destruction to rocks in its path. Oftentimes the disintegration of the parent material is so extensive that rivers heavily laden with rock detritus flow thick with the hue and tint of the mineral particles. As the velocity of the rivers slow, mineral particles begin to settle and raise the floor of the waterway. This activity eventually causes flooding and/or changes in the river's course and, after centuries, forms the base of the nutrient rich river valley soils.

Another ill effect on rock formations is frost. When water freezes it expands with a force equivalent to 150 tons of pressure per square foot. Water that seeps into the cracks and/or crevices of rocks in the late Fall, Winter or early Spring will inevitably fracture them. If a broken piece is heavy enough and falls from high enough, gravity will assist as the boulder comes crashing down, chipping away at itself and much of what is in its path.

Porous rock surfaces that can absorb water will literally explode (in very slow motion) from the force of frost. What remains is often what appears to be dust depending on how extensive the water absorption was to begin with. This dust is then carried off by wind and/or water currents to join in the soil formation process.

Wind is another very effective force that wears away at the surface of rock. As more rock particles are liberated and become airborne, the wind becomes somewhat of a sandblaster to hasten its weathering effect even further. The wind also carries micro-organisms that chemically weather rock (mentioned later in this chapter).

Roots of plants have both a physical and chemical weathering effect on rocks. Fine root hairs that find their way into cracks or crevasses of rock surfaces eventually grow and expand, oftentimes fracturing large boulders into smaller pieces or liberating various sized pieces from rock cliffs or mountains. If a root is only able to create an avenue through the rock for water to seep into, it has done enough to begin weathering.

The chemical weathering of rocks occurs from acids and enzymes created by organisms that live in the root zone and by the roots themselves. Inorganic acids formed by soil chemical reactions also play a major role in the weathering process.

The cycle of glaciation has a tremendous weathering effect on parent material. The unimaginable weight of a glacier grinding rock against rock creates immense quantities of glacial dust that contribute to the structure and fertility of the soil. After a glacier recedes, the rock dust is distributed over a broad area by wind and water currents. The need for re-glaciation to replenish the mineral content of soils is a relatively popular notion.

Trespass is another very effective weathering force in nature. Animals, ranging from the elk to the earthworms, crumble, crush, break, and wear down rock surfaces into soil components. The combined forces of trespass and gravity can have a synergistic effect when animals create rock slides of different magnitudes while traveling up or down steep slopes. The decomposition of excrement left by trespassers adds another chemical weathering effect by feeding the biological creation of organic acids that can corrode rock surfaces.

Tectonic movement is another force to be considered in soil formation. Although the earth's surface is relatively stable compared to prehistoric eras when earthquakes, volcanic eruptions and continental plate movement was peaking, there is still enough movement in the earth's surface to cause considerable damage to rock formations. Very small tremors can loosen and crush a significant amount of parent material.

CHEMICAL WEATHERING

Surprisingly, the largest facility on earth where chemical reactions occur is the natural environment. Elements that occur naturally react with each other on a regular basis forming compounds, many of which react with other chemical elements or compounds. This constant manufacture and disintegration of chemicals in nature is an integral part of terrestrial functions. One of effects of this natural chemical activity is the weathering of rocks into soil.

Water is chemically expressed as H_2O. Those two elements (i.e. oxygen and hydrogen) can chemically react with many elements in nature. The entire water molecule can react in a process called *Hydration*. Rock minerals that bond with one or more water molecules become hydrated and are more easily dissolved into a soil system (e.g. hydrated lime). *Hydrolysis* occurs when the hydrogen atom derived from water bonds with mineral(s), oftentimes forming acids that can further weather rock surfaces. Chemical chain reactions initiated by either the hydrogen or the oxygen in water can change the original composition of rock into mineral compounds with completely different structures and reactive characteristics within the soil.

The formation of Inorganic acids, such as sulfuric, or hydrochloric, occur naturally through reactions between different soil compounds and water. These acids are extremely effective at separating and dissolving rock components. Carbonic acid, another effective weathering agent, is formed from the combination of carbon dioxide and water, two relatively abundant substances in the soil.

Hundreds of different organic acids are formed in nature by plants, animals and microbes. These include citric, acetic, amino, lactic, salicylic, tannic, nucleic and humic to name a few. All of these acids have varying abilities to react with rock surfaces and liberate mineral from the parent material.

Microbes can dissolve mineral from rock in association with three basic functions: *Assimilation*, the creation of *organic acids* and the creation of *inorganic acids*. An illustration of this is Mycorrhizae, a family of fungi that can dissolve mineral nutrient from rock surfaces. These organisms work under the soil's surface in a symbiotic relationship with the roots of trees and other perennial plants. The fungi attach themselves to roots and, in exchange for a small amount of carbohydrate supplied by the plant, they give water and mineral nutrient back which they have retrieved and transported via their hyphae from soil depths.

Assimilation is the direct need for mineral nutrient by the organism. Lichen, for example, are symbiotic combinations of fungi and algae that can attach themselves to rock surfaces and dislocate minerals for its own nutrition (see figure 1-1). These organisms can be observed around the world and in some of the harshest climatic conditions. The microscopic hyphae of the fungi component can

Figure 1-1: Lichen growing on rocks

actually penetrate the finest cleaves or cracks of certain rock releasing enzymes and acids powerful enough to liberate mineral from surrounding rock surfaces. Lichen also exert a constant pulling action on the surface they attach themselves to, physically separating particles from the parent material. Eventually, the minerals assimilated by Lichen and similar organisms are re-cycled through the agency of other organisms that either feed on lichen or decompose their remains after death. However, now the nutrients are in an organic form which is easier for nature to process. Unless these elements are removed from a given area (by cropping, erosion or other means) they tend to accumulate over time to create fertility for both plants and the other organisms that live there.

The creation of organic acids by most organisms is either for assimilation (mentioned above) or as by-products of their metabolism. Regardless of the purpose, these acids contribute immensely to the formation of soil. Since organic acids can be found almost anywhere microbes are found, their weathering activities are not restricted to arable land. They also work on mountains, deserts, under glaciers and even on man made objects such as buildings and

INFLUENCE OF ACIDITY CREATED BY AZOTOBACTER ON SOLUTIONS OF Ca, Mg, and K FROM PARENT MATERIAL

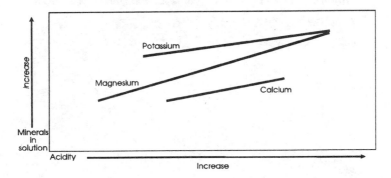

Figure 1-2 (Waksman 1936)

bridges. Figure 1-2 shows the influence of organic acids created by azotobacter bacteria on mineral content of soil solutions. These beneficial bacteria also fix nitrogen from the atmosphere.

The creation of inorganic acid such as sulfuric and nitric, is formed in the soil as an *indirect* result of microbial activity. Specific organisms such as sulfur bacteria or nitrifying bacteria oxidize (combine with oxygen) sulfur and nitrogen which can react with water to form strong inorganic acids. The chemical reaction between rock surfaces and strong acids changes both substances. The rock obviously disintegrates and the acid can then become a mineral salt such as potassium sulfate that is soluble and more available to plants and other soil organisms.

Acid rain created by industry is a recent phenomena that also causes chemical weathering of rocks. The effects of this type of precipitation containing both sulfuric and nitric acids can be seen on limestone and marble edifices around the world.

The chemical weathering effect that roots have is from several sources. Carbon dioxide given off by root hairs can combine with water to form carbonic acid. Research has shown significant mineral release from carbon dioxide when dissolved in water. This process is known as carbonation and involves the combining of mineral ions such as potassium, magnesium or calcium to a carbonate ion forming a mineral salt. Approximately 20% of the carbon that plants fix from atmospheric carbon dioxide is exuded from the roots as carbon compounds, many of which are, or are converted to organic acids.

Plant roots also have an indirect participation in the formation of organic acids because of their relationship with soil micro-organisms. The population of microbes within the rhizosphere (i.e. the soil region immediately surrounding the root hair) is always significantly higher than in other parts of the soil (see figure 1-3). The production of organic acids and enzymes that are destructive to rock is proportional to the microbial population.

INFLUENCE OF PLANT ROOT
ON BACTERIA POPULATIONS

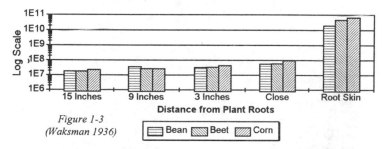

Figure 1-3
(Waksman 1936)

The different types of parent material determines the rate at which rock is weathered and, to a large extent, the different sized particles found in soil. Limestone, for example, is a rock that is easily weathered and can eventually dissolve so completely that very few particles can be found. Quartz, on the other hand, weathers very slowly and because of its molecular structure seldom is reduced to a size smaller than that of sand (see table 1-1).

TABLE 1-1

CLASSIFICATION	PARTICLE DIAMETER (mm)
Clay	> to 0.002
Silt	0.002 to 0.05
Sand	0.05 to 2.0
Gravel	>2.0

Surface area is another factor that has an influence on weathering. The greater the surface area exposed to soil acids, the faster the rock can be dissolved. A fist sized stone may have several square inches of surface area when intact, but when ground into a fine powder the

Figure 1-4 (Scanning electron micrographs courtesy
Dr. Bruce F. Bohor, Illinois State Geological survey)

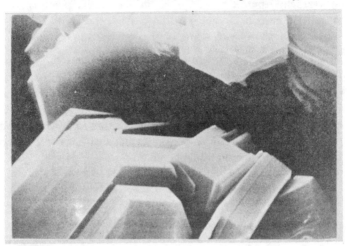

overall surface area increases to several acres. Conversely, the
amount of time it takes for nature to dissolve the material would be
measured in years for the powder as opposed to centuries for the
intact stone.

Rocks comprised of predominately aluminum, potassium or
magnesium silicates (generally insoluble compounds) are commonly
weathered into very small plate-like particles classified as clay (see

figure 1-4). Clay particles, because of complex substitutions of elements within their molecular structure, inherently have a negative electro-magnetic charge which enables them to adsorb positively charged ions (cations) such as potassium (K), calcium (Ca) or Magnesium (Mg) (see figure 1-5) (see chapter 7). This magnetic ability is described as colloidal (from the Greek *koll* meaning glue and *oid* meaning like; *see glossary*) and is inherent in humus particles as well. Because of the nature of colloidal particles in the soil its presence is crucial to the entire soil system and to plant growth (See chapter 7).

CLAY PARTICLES
MICELLES

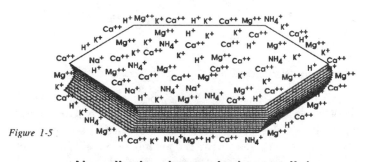

Figure 1-5

**Negatively charged clay particles
shown with typical plate-like
appearance and swarm of adsorbed cations**

Soils formed from rock are called mineral soils and they are the most common type of soil on the earth's crust. An analysis of a well developed mineral soil might contain around 90% rock particles on a dry basis (see figure 1-6). Volumetrically, 50% of this type soil will be made up of air and water and if it is a rich, healthy soil, an average of only 5% will be organic matter.

There are ninety elements that are naturally occurring on earth and most are found (at least in trace amounts) almost everywhere. However, the most common elements found in the mineral component of soil are oxygen, silicon, aluminum, calcium, sodium, iron, potassium and magnesium (see figure 1-7). The oxygen that exists in mineral is part of its chemical structure and not in a gaseous state.

SOIL COMPONENTS
Typical analysis of a well developed loam

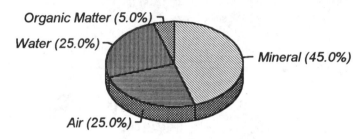

Organic Matter (5.0%)
Water (25.0%)
Mineral (45.0%)
Air (25.0%)

Figure 1-6

Many mineral elements exist in a molecule with oxygen.

All of these weathering forces, both physical and chemical, combine to form soil particles from rock that are better known as sand, silt and clay. These particles are defined by size as shown in Table 1-1. Most soils have a combination of all three sizes of particles and are classified differently depending on the percentage of each (see chapter 7).

ORGANIC MATTER

As minerals from rock weathered into available nutrients, organisms that could utilize them, such as plants and microbes began to evolve and to create biomass. The proliferation of these organisms depended on the quantity and quality of the mineral resources available from their individual environment. As this new life cycled into death, new organisms evolved that utilized the nutrients contained in the dead tissue left behind. The system that evolved created a process whereby mineral nutrients could be recycled almost indefinitely. The accumulation of organic matter is a complex subject and is discussed in chapter 2. However, it is important to note here the characteristics of organic matter (OM) that physically hold the soil together. OM increases the life span of any given soil by

MINERAL COMPONENTS
8 Elements Comprise 98% of All Soil Mineral

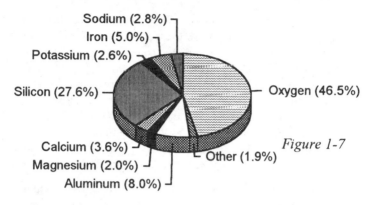

Sodium (2.8%)
Iron (5.0%)
Potassium (2.6%)
Silicon (27.6%)
Oxygen (46.5%)
Calcium (3.6%)
Other (1.9%)
Magnesium (2.0%)
Aluminum (8.0%)

Figure 1-7

inhibiting those weathering forces (especially wind and water currents) from destroying it beyond its most productive stage.

WEATHERED TO DEATH

Soils are subject to eventual ruination by the same forces that created them. The inherent mineral content of any soil may be vast but it is also finite and can eventually be depleted by weathering. At some point, perhaps a billion years from the time of a soil's beginnings, leaching and/or erosion can eventually deplete the mineral resources that are crucial to the existence of living organisms. As the availability of minerals wanes so does the biomass that sustains the level of native organic matter which, structurally, protects the soil from erosion and stores much of the moisture and nutrients necessary for biotic development. As the depletion continues, only the most weather resistant particles of quartz and other mineral compounds remain. The soil can eventually evolve into a habitat that sustains only the hardiest of organisms.

Erosion, one of the forces that weathered rock and carried particles of parent material to deposits called soil, can also carry it away. Oftentimes it can be carried all the way into an ocean where tremendous pressures for long periods of time can transform the mineral particles into sedimentary rock. Perhaps, a million years from now, a broad tectonic event will lift the recreated rock into a mountain and the process can begin all over again.

PLANTS

Plants are the connection. They are the link between the atmosphere and the pedosphere (heaven and earth). Their remains make rich soil from weathered rocks, and they provide nutrients and energy for most other living things on earth. They complete the cycles of water, nutrients and energy. Plants are the producers. They combine the energy from the sun, the carbon dioxide in the atmosphere and the mineral nutrient in the soil to synthesize sustenance either directly or indirectly for almost every other living thing on earth. All of the protein, energy and carbohydrates that humans need come from plants, or from animals raised on plants.

As prehistoric soils formed, plant life slowly began to appear. At first very small, perhaps microscopic, plants were all that could get established because of the meager amounts of available nutrients coupled with a harsh environment. As time went on, the accumulation of humus from the remains of countless generations of micro-plant-life slowly enriched the soil, creating an environment that could support a larger and more diverse population. The cycle snowballs for a period of time creating a richer soil with every successive increase in vegetative establishment until an ecological equilibrium is reached. At this point the production of organic matter by plants is roughly equal to the environmental factors that destroy it.

The evolution of photosynthetic organisms such as plants changed the ecosystem of the planet dramatically. Prior to these autotrophs (i.e. producers), organisms existed on available energy. When the energy was depleted, the organisms became extinct and new organisms evolved to recycle the energy left behind by its predecessors. The extreme volatility of the ecosystem was stabilized by photosynthetic organisms that essentially bottled the energy of the sun and made it available to other living things. The renewability of energy on earth enabled many species of heterotrophs (i.e. consumers such as mammals) to exist for many generations without depleting their food supply. This phenomenon allowed for the systematic evolution of different species into present day life forms.

Land plants are composed of two fundamental parts: the root system which anchors the plant and absorbs water and mineral nutrients, and the shoot system which consists of its trunk, branches, stems and leaves. Those parts of the plant that connect the roots to the leaves provide transport passageways for water and nutrients, and structural support. The surface of leaves contains pores called *stomata* that function as a gas exchanger and part of the plant's ventilation system.

The plant's link to the atmosphere is its use of carbon dioxide (CO_2) gas to create almost all the carbon compounds that exist on earth (with the exception of minerals such as calcite or dolomite). Without CO_2 plants could not live and neither could any of the organisms that depend on the nutrients that plants produce. The balance of CO_2 with the other gases in the atmosphere depends largely on plants or, more accurately, the utilization of energy from plants. Whenever energy is extracted from plant tissue (or animal tissue), whether by animals, microorganisms or humans, CO_2 is given off. Burning fossil fuel, wood or other organically derived carbon is another extraction of energy that gives off CO_2. A quantitative or even qualitative change in the plant population on earth can have an effect on the balance of CO_2 in the atmosphere which, in turn, can affect other edaphic factors such as the climate. Conversely, atmospheric changes affecting CO_2 content caused by events such as the burning of fossil fuel or volcanic activity can influence both climate and plants.

Plant roots make the inorganic/organic connection in the soil. The fine root hairs combine so well with the earth's minerals; it is difficult to tell exactly where the soil ends and the root begins. The roots of plants depend on the shoot system for nourishment. Throughout the plant's life proteins, carbohydrates and other organic nutrients flow down to the root system from the leaves where energy from the sun, carbon dioxide from the atmosphere and minerals drawn from the soil combine in a complicated process called *photosynthesis*.

The root system and plant tops act like different organisms living together symbiotically. The leaves are autotrophic, utilizing carbon dioxide from the atmosphere to manufacture organic carbon compounds and giving off oxygen as a waste product. The root systems act like a heterotrophic organism, dependent upon the sugars and other compounds produced in the leaves and utilizing oxygen from the soil while giving off carbon dioxide as its waste.

The roots of plants are more than just anchors and siphon tubes extracting water and minerals from the soil. The rhizosphere (i.e. the zone of close proximity around each root hair) is teeming with and supporting life in the form of billions of microorganisms. Approximately 20% of the photosynthesized nutrients from plant leaves are exuded from the roots into the rhizosphere where it supplies nourishment to these organisms. The activity of these organisms help the plant in numerous ways:

-mineralize organic nutrients
-compete antagonistically with pathogens
-weather parent material
-decay plant residues into humus and return CO_2 to the atmosphere
-mobilize water and nutrients (i.e. mycorrhizae)
-(some) fix nitrogen from the atmosphere

Soil factors that influence plant growth are also affected by plants. Different environments such as prairies and forests are initially formed by other factors, e.g. climate, parent material and topography. But the establishment of certain types of plants soon becomes another factor of the environment. The plant factor influences the establishment of different types of organisms, changes the effects of topography, and can even create subtle differences in the atmosphere that can affect the climate. Different environments are eventually stabilized by plants, completing the cycles of nutrients, cycles of water, and the life cycles of organisms dependent on plants for nutrients. These environments are still changing, but the plantlife acts as somewhat of a parachute, stabilizing and oftentimes removing much of the volatility from their evolution.

Plants protect the soil environment by providing shade to slow down the evaporation of precious moisture needed for its own life and the lives of other soil organisms. Plants also control erosion with their root systems and the organic matter they produce. The adhesive qualities of humus hold soil particles together and provide greater absorption of water through an increase in soil porosity. This increase in water absorption provides a relative decrease in surface run off.

The mechanism by which plants produce the proteins, sugars, starches, fat and fiber needed by the consuming species of the world is chemically quite complex. The plant's ability to utilize the endless energy from the sun and the small percentage of carbon dioxide in the earth's atmosphere to produce all of these organic nutrients is nothing short of a miracle.

In simple terms, there exist small openings called stomata, mostly on the underside of the plants' leaves, that allow the transference of gas and moisture into and out of the plant. Carbon dioxide, much of it generated by the microbial decay of organic residues in the soil, is intercepted by plant leaves as it escapes the soil. Plant cells that contain chlorophyll are excited by the energy of the sun and conduct chemical reactions within the leaves that combines the carbon and some of the oxygen from carbon dioxide (CO_2) with hydrogen from water, nitrogen, phosphorus, sulfur and other minerals from the soil to form the complex organic compounds that give life to the plant, to its roots and to every other organism that depends either directly or indirectly on plants for their sustenance. These six elements mentioned above make up 99% of all living matter.

Other nutrients considered essential to plants are:

calcium	molybdenum	magnesium	boron
sulfur	copper iron	cobalt	zinc
chlorine	potassium	manganese	

All of the carbon plants use is derived from the atmosphere. Oxygen comes to the plant from both the atmosphere (as CO_2) and

PLANT NUTRIENT NEEDS

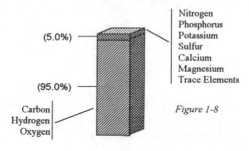

(5.0%) — Nitrogen Phosphorus Potassium Sulfur Calcium Magnesium Trace Elements

(95.0%) —

Carbon Hydrogen Oxygen

Figure 1-8

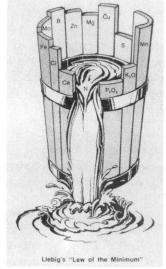

Liebig's "Law of the Minimum"

White 1982

water (H_2O) which also delivers hydrogen. These three elements constitute the basic building blocks of all organic compounds and comprise approximately 95% of a plant's diet. The remaining 5% is in mineral form and derived from the soil (see figure 1-8).

In 1840 Justis von Liebig discovered that plants utilized nutrients from the soil in a mineral form and that each essential nutrient could limit the growth of a plant even if only a trace amount was needed. From this discovery came Leibig's Law of Limiting Factors (see figure 1-9).

Both the direct and indirect effect that plants have had, and are still having on the earth's evolution, is profound. The existence of almost all living species is both controlled by and controls the worldwide population and diversity of plantlife. But of all the different species of heterotrophs affected by plants, the influence of the human race has created the greatest amount of change in the shortest period of earth's history.

THE HUMAN FACTOR

"As most of the productive land on Earth has been taken over for human use and more or less drastically altered in its biotic character, the remaining virgin forests, grasslands, and other natural areas throughout the world increasingly are islands in a vast sea of human disturbance." -Ehrlich 1991

In the past ten thousand years of Earth's history a new factor in soil formation occurred through the development of agriculture. The establishment of the human species as farmers exerted a powerful influence on the evolution of soil.

Agriculture began after two million years of hunting and gathering. It is not exactly clear why early man adopted cultivation as a means of survival because it was not necessarily easier, but the development of agriculture and all the organization associated with it was, by most definitions, the beginning of civilization. However, population growth was, and still is, a side effect of this life-style, and the ecological toll of extracting nourishment from the soil for ever-increasing numbers of people has created wastelands from what were once productive soils.

These ecological disaster zones are a result of some of the following anthropogenic circumstances:

Erosion - Erosion is a significant disturbance to the soil's state of being. The same currents of wind and water that brought the soil from its original parent material can also carry it away if it is not anchored by plants and glued together with humus. Clear cutting of forests and bare ground cultivation leave the topsoil vulnerable to tremendous losses from erosion. One rain and/or wind storm can remove what took millions of years to accumulate if the soil is not protected by plants and humus. Silt, a by-product of erosion, can eventually be deposited into lowland soils changing their ability to percolate moisture and their overall porosity.

Destruction of humus - Stable organic matter is considered to be the wealth of a soil because of its abilities to hold water and nutrients, and its ability to support an incredible diversity of soil life. Yet, conventional cultivation squanders this asset away. Constant aeration and irrigation, over-liming, over-fertilizing and exposing bare ground to the heat of the sun all contribute to the depletion of organic matter (see chapter 2 and chapter 5).

Water Pollution - The pollution of the world's water, whether fresh or salt, open or underground, will eventually take its toll on the earth's inhabitants. Given the value of water to the existence of all living things, its demise will cause ecological disasters on a scale beyond compare to anything in recorded history. The effect on soil from water pollution is somewhat of a moot point if the availability of clean water is life's most limiting factor. It is also a subject that would require volumes of work to even outline given the infinite number of variables on earth affected by water (See chapter 3).

Destruction of ecosystems - Slash and burn agriculture in the fragile and diverse rain forests inflicts a painful and perhaps permanent wound on earth ecology. Changes in global climates and water cycles caused by altered plant transpiration and soil respiration have a profound effect on soil characteristics and evolution. There is also the systematic elimination of countless thousands of different species of plants, animals and other organisms that may have either a direct or indirect influence on soil. Ironically, many of these organisms being pushed into extinction might have been the raw materials needed for the development of new medicines or other advances in research. As much as forty percent of the different species of organisms live in these tropical humid environments that occupy less than seven percent of the earth's land area. There is no doubt that many of these species have an intimate ecological link to the quantity, quality and characteristics of soil all over the planet.

From a regional perspective, the denuding of any environment leaves soil without protection from the sun, wind and water. Excessive heat from the sun accelerates the decomposition of soil organic matter making the soil even less resistant to wind and water erosion. Since forest fertility is usually restricted to a thin layer of topsoil, its potential for regeneration is increasingly lessened the longer the soil is exposed.

Waste stream - During the era of hunting and gathering the human species, along with most other organisms, had a natural and efficient method of dealing with wastes. This inadvertent recycling enriched the environments where our ancestors tread, giving back some of what was taken. Today, our waste stream seems to flow down a dead end street (for the most part), accumulating into what has become one of the most pressing problems on earth (waste management). Regardless of how this problem evolved or how it will be solved, the fact remains that everything in our waste pile was once a natural resource that will never be returned to usefulness or to its natural state. Many of these resources are minerals and organic compounds derived directly or indirectly from the soil. The linear, non-cycling direction of these compounds may eventually deplete the soil to a point where no more wastes can be produced. The implications of this scenario also eliminates the intermediate phases between production and so-called uselessness, including human needs.

Greenhouse warming - In Greek mythology, the titan Prometheus, who created man from clay, stole fire from the gods to give to mankind for warmth, cooking and the forging of tools and weapons. His actions angered Zeus, the father of the gods, who punished Prometheus with a horrible eternity of pain and bondage. Zeus initially felt threatened by humans with the power of fire and was compelled to annihilate all mortals by creating a great conflagration on earth but, upon reconsideration, thought it unnecessary to take any action because human beings would eventually use fire to destroy themselves.

Although mythical, the use of fire by humans has shown the complacency of Zeus to be valid. When nature introduced fire to man, it quickly became a tool for hunting, by flushing animals out of the thick forests. It is credited with the extinction of several ancient species of mammals and probably many other organisms that slipped through the cracks of fossil history. However, another problem from the era of fire has been the tremendous amount of carbon dioxide (CO_2) being returned to the atmosphere. In chapters 2 & 5 we see that the cycle of CO_2 from soil into the atmosphere is essential for the growth of plants and all of the life that depend upon them. However, the accumulation of excess CO_2 can insulate the earth from escaping heat, causing greenhouse warming. Thousands of years of burning fossil fuels for hunting, clearing land, cooking, warmth and industry have increased the CO_2 content of the atmosphere by about twenty five percent. The gradual increase in temperature around the world can and will have a profound effect on soil evolution. Not only will global warming reduce native levels of organic matter (see chapter 5) but it can also affect precipitation, frost activity and habitability of different environments to organisms, all of which are integral components of soil formation. Regional droughts are likely to be more frequent, longer lasting and more severe with warmer global conditions, increasing the likelihood of wind or water erosion that can remove a significant amount of fertile topsoil.

Man's use of energy - Fossil fuels, the source of most of the energy utilized by civilization, is derived from biological synthesis in the soil. Reservoirs of organic energy that took hundreds of millions of years to accumulate are being depleted in a matter of centuries. The amount of fossil energy utilized to cultivate and harvest an acre of soil in this modern era of mechanized agriculture is often greater than the energy produced by the crop. This net negative production cannot go on indefinitely.

Fertilizers and pesticides - At the dawn of the so-called "Green Revolution", artificial fertilizers were thought to be the proverbial gift of the gods, giving sustainability and increased yields to every acre of soil on earth. However, in less than fifty years the magic of processed inorganic fertilizers began to fizzle and is now being blamed for as many problems as it was credited for gains. Many of those problems are being *managed* with applications of pesticides to increase yields and the marketability of agricultural products. These chemicals seep into the ecosystem, altering the balance of organisms that play direct or indirect roles in soil formation and maintenance.

Eye ball society - Civilization, especially in the U.S., is strongly attracted to products that look good. The *If-it-looks-good,-it-is-good* mentality has created agricultural markets that require perfect looking food. Unfortunately, the price we pay for this blemish-free presentation may be an indelible disruption of the ecosystem from the pesticides and other toxins needed to produce it. The irony is that we would rather eat food laced with toxins than share some of it with another organism. The indirect effect of this chemical warfare on non-target organisms can not be accurately predicted. However, soil is almost always affected by changes in the biological environment.

Irrigation - The development of arid or semi-arid lands for agriculture by irrigation have, in many cases, created salt deserts where very little can grow. Unlike rain, water from underground aquifers contain natural mineral salts. These salts can accumulate at the soil's surface when this water is used for irrigation. If annual precipitation cannot leach the salts from the surface, as is the case in arid and semi-arid regions, the accumulation can reach phytotoxic levels. Drained aquifers is another consequence. These circumstances are already occurring in many places. Farm land surrounding the Aral Sea in the (formerly) U.S.S.R. have been completely decimated by pumping sea water for irrigation. Not only has the soil become a salt desert as far as the eye can see, but the sea itself has been drained far from its original shoreline, and the chemical effluent from years of fertilizer and pesticide applications have laid dormant the sea's once rich fishing industry. Some of the more productive valleys in California are fast approaching the fate of the Aral Basin from the irrigation of soils that would not naturally support crops.

Seed stock - Genetic diversity is another factor of soil evolution. As crop scientists narrow the seed stock of the world down to less diverse varieties, the major crops of the planet become more susceptible to massive failure from either drought, pestilence or some other environmental stress condition. The repercussions of such a scenario could spell disaster to the human race which depends so heavily on so few varieties. That same susceptibility must also be manifested in the soil. The plant-soil interactions of a wild environment, such as a forest, produce a soil rich in both biotic and abiotic diversity. The consequences of limiting plant varieties to only a few can eventually cause changes in soil structure, biota or other factors of soil evolution.

Acid rain - Industry's use of coal as fuel, especially soft coal, has increased the amount of sulfur and other contaminants in the atmosphere that, when combined with moisture can form strong inorganic acids. Earlier in this chapter, the important role that both organic and inorganic acids play in the formation and degradation of soil, was discussed. Acids from the sky accelerate this process and change the pH of waterways that are constantly reacting with the parent material inherent in different environments. The latter part of this phenomenon may not necessarily be adverse because the fresh mineral from the parent material may replace what is leached out of the soil. However, with modern flood control systems, it is more likely that this valuable detritus will be washed out to sea.

The soil is also affected indirectly by the destruction of millions of trees from acid rain. These organisms can no longer produce organic matter, hold and shade the soil, or extract CO_2 from the atmosphere. Their roots no longer interact with soil organisms or minerals and all other relationships with the environment are terminated. The evisceration of fish colonies from the pH changes in lakes and streams caused by acid rain can also affect soils. This dramatic change in the food web ripples through every aspect of the ecosystem.

Ozone depletion - The destruction of ozone in the upper atmosphere by artificial gases such as CFC's (Chloroflorocarbons) changes the intensity of the earth's energy source, i.e. the sun. Ninety nine percent of the sun's ultraviolet (UV) rays are filtered by upper atmospheric ozone. A one percent depletion of ozone will double the amount of UV that reaches the earth. Factors of soil formation such as climate and living organisms can be significantly altered by changes in sunlight's color spectrum or levels of radiation from the sun. Not much research has been done regarding the influence of ozone depletion on soil (probably because of the large number of variables involved), but in accordance with edaphic theory, *energy that affects anything in nature will eventually affect everything else*.

Desertification - The occurrence of drought in certain regions of the world is a natural phenomenon that is usually cyclical. Unfortunately, in most of these arid and semi-arid environments it is also unpredictable. During periods of history when rains are adequate or even abundant, the resulting vegetation can support a growing population. However, when the unforeseen drought occurs, the already stressed vegetation is often pushed to extinction by people and animals who are struggling to survive. By the time rains return to the region, significantly fewer of the perennial indigenous plants can be revived, and the soil's vulnerability to erosion is dramatically increased, creating an area that is even less suitable for raising crops or livestock.

Transport systems - As today's farms become situated farther away from its markets and processing centers, there is a growing interdependence on transport systems to move food to the population and inputs such as fertilizer, equipment or pesticides to the farm. The millions of miles of roads needed for relocation isolates the soil and it organisms from the atmosphere and the climate. In addition, the vehicles needed for the movement of this material deplete energy reserves and create atmospheric pollution. All of these changes have a direct or indirect effect on the soil.

Mining - The mining of non-renewable resources for energy or other raw materials disrupts the soils ecosystem to such an extent that it takes generations for biotic equilibrium to be reestablished, if it is even possible for recovery to occur. The mining of renewable resources can be equally devastating. The forests around the world, whether tropical or temperate, regulate global weather systems that have a direct effect on soil. As more of these ecosystems are altered by clear cutting for timber and pulp, there is a domino effect that alters all of the ecologically linked earth components, including soil.

Economic pressure - Many farmers trying to scratch a living out of the soil are economically stressed by debt and other expenses of a modern society. Their dilemma strains the resources of the soil as greater production is forced from their land. Maintaining or improving soil quality is simply a luxury they cannot afford.

SUMMARY

The basic knowledge of most organisms on earth is limited to survival techniques. The exception, of course, is the human being who has learned to learn and can reason beyond instinct. Unfortunately, no operating manual came with this planet and instruction has always been a trial and error affair. If mistakes are made on a large scale, the ecological reaction can be devastating. Unfortunately, as humans have acquired more information, they have also caused more disturbance to the soil. H.G. Wells described the situation best in his passage that alludes to the history of humanity as a race between learning and disaster.

The individual human life spans less than a blink in the time frame in earth's evolution. Short term thinking comes naturally to us. However, the time has come for civilization to abandon the one, five or ten year plans and to adopt a 1,000, 5,000 or 10,000 year plan if the life of the earth is going to survive the mortality of mankind.

Sources:

Arshad, M.A. and Coen, G.M. 1992, Characterization of soil quality: Physical and chemical criteria. American Journal of Alternative Agriculture v7 #1 and 2, 1992 pp 25-31. Institute for Alternative Agriculture, Greenbelt, MD

ASA# 47. 1979, Microbial - Plant Interactions. American Society of Agronomy. Madison, WI

Bear, F.E. 1924, Soils and Fertilizers. John Wiley and Sons, Inc. New York, NY

Brady, N.C. 1974, The Nature and Properties of soils. MacMillan Publishing Co. Inc. New York, NY

Brown, B. and Morgan, L. 1990, The Miracle Planet. W. H. Smith Publishers, Inc. New York, NY

Brown, H. Cook, R. and Gabel, M. 1976, Environmental Design Science Primer. Earth Metabolic Design. New Haven, CT

Ehrlich, P.R. and Ehrlich, A.H. 1990, The Population Explosion. Simon and Schuster. New York, NY

Ehrlich, P.R. and Ehrlich, A.H. 1981, Extinction: The Causes and Consequences of the Dissappearance of Species. Random House, Inc. New York, NY

Fowler, C. and Mooney, P. 1990, Shattering: Food, Politics, and the loss of Genetic Diversity. The University of Arizona Press. Tucson, AR

Greulach, V. A. 1968, Botany Made Simple. Doubleday & Company, Inc. Garden City, NY

Hamaker, J.D. and Weaver, D.A. 1982, The Survival of Civilization. Hamaker - Weaver Publishers. Michigan, CA

Hillel, D.J., 1991, Out of the Earth: Civilization and the Life of the Soil. The Free Press. New York, NY

Huang, P.M. and M. Schnitzer 1986, Interactions of Soil Minerals with Natural Organics and Microbes. Soil Science Society of America, Inc. Madison, WI

Jenny, H. 1941, Factors of Soil Formation. McGraw - Hill Book Co. New York, NY

Kahn, H., Brown, W. and Martel, L. 1976, The Next 200 Years. William Morrow & Co. New York, NY

Karlen, D.L., Eash, N.S. and Unger, P.W. 1992, Soil and crop management effects on soil quality indicators. American Journal of Alternative Agriculture v7 #1 and 2, 1992 pp 48-55. Institute for Alternative Agriculture, Greenbelt, MD

Logan, W. B. 1992, Hans Jenny at the Pygmy Forest. Orion v 11 #2 Spring 1992 pp 17-29. Myrin Institute. N.Y., NY

Parnes, R. Fertile Soil: A growers Guide to Organic & Inorganic Fertilizers. Ag Access, Davis, CA

Parr, J.F., Papendick, R.I., Hornick, S.B. and Meyer, R.E. 1992, Soil Quality: Attributes and relationship to alternative and sustainable agriculture. American Journal of Alternative Agriculture v7 #1 and 2, 1992 pp 5-10. Institute for Alternative Agriculture, Greenbelt, MD

Ray, P.M. 1972, The Living Plant. Holt, Rinehart and Winston, Inc. N.Y., NY

Raven, P.H. and Curtis, H. 1970 Biology of Plants. Worth Publishers, Inc. New York, NY

Richards, B. 1993, Personal communication. Newbury, VT

Sagan, D. and Margolis, L. 1988, Garden of Microbial delights: A Practicle Guide to the Subvisable World. Harcourt Brace Jovanovich, Publishers. Boston, MA

Smith, G.E. 1942, Sanborn Field: Fifty Years of Field Experiments with Crop Rotations, Manures and Fertilizers. University of Missouri Bulletin #458. Columbia, MO

SSSA# 19. 1987, Soil Fertility and Organic Matter as Critical Components of Production Systems. Soil Science Society of America, Inc. Madison, WI

Stork, N.E. and Eggleton, P. 1992, Invertebrates as determinants and indicators of soil quality. American Journal of Alternative

Agriculture v7 #1 and 2, 1992 pp 38-47. Institute for Alternative Agriculture, Greenbelt, MD

Stuart, K. 1992, A Life With The Soil. Orion v 11 #2 Spring 1992 pp 17-29. Myrin Institute. N.Y., NY

Wallace, A., Wallace, G.A. and Jong, W.C. 1990. Soil Organic Matter and the Global Carbon Cycle. Journal of Plant Nutrition 1990 v13 (3/4): 459-456

White, W.C. and Collins, D.N. (Editors) 1982, The Fertilizer Handbook. The Fertilizer Institute. Washington, DC

Wilson, C.L., Loomis, W.E. and Steeves, T.A. 1971, Botany. Holt, Rinehart and Winston. New York, NY

Chapter 2

HUMUS:
A STABLE SOIL ORGANIC MATTER

Humus, like so many materials on earth, is abundant, renewable and essential for life to exist on this planet. However, humus is complex, and even after hundreds of years of research, no one really knows exactly what it is.

The term *Humus* doesn't really describe anything specific. It is like using the word *dog* to describe a German Short Haired Pointer or a Russian Wolf Hound. Humus is sometimes defined as the end product from the decomposition of organic residues. But since it never remains in a static condition, it is hard to refer to humus as an end product. Furthermore, the composition of humus in one soil can be so structurally, chemically and visibly different from humus in another soil that it is difficult to refer to them both as the same thing.

The popular scientific definition of humus is *A more or less biologically stable, dark, amorphous material formed by the microbial decomposition of plant and animal residues.* It is difficult to visibly differentiate humus from organic matter in other stages of decay. Compost, well rotted manures and peat are not necessarily humus. However, at some hard-to-define point, all of these organic materials will contribute immeasurably to the humus content in the soil.

Over the years, much information has been gathered about humus. Certain components have been identified. Its nature and its properties are fairly well known, and the factors that control its existence are pretty much accepted as common knowledge. However, to date, an indisputably accurate method of extracting humus from soil has yet to be discovered, which severely limits the study of this material.

Attempts to define humus date back to the time of the Roman empire, but only in 1761 was humus first linked to the decomposition of organic matter by J. G. Wallerius. In that era it was thought that plants were able to derive nutrients directly from humus however, in 1840, Justus von Liebig discovered that plants can only assimilate soil nutrients in an inorganic form, and that plant food must be changed into a mineral form first. Liebig believed that this occurred from chemical reactions in the soil. About twenty five years later, attention was called to the role of microorganisms in the mineralization of nutrients from humus. It was in the early 1900's when most of the significant research on humus occurred. However, much of the information discovered became somewhat obsolete at the dawn of the chemo-agricultural age in the early 1940's when synthetic mineral nutrient seemed to answer every farmer's dream.

ORIGINS

The formation of humus begins when the residues from plants and animals come in contact with microbial life in the soil. Much of the carbon compounds contained in those residues are proteins, carbohydrates and energy for the various bacteria, fungi and actinomycetes involved in the decay process.

Aerobic microorganisms are the most adept at decomposing organic matter. They need an environment where there is an adequate amount of free oxygen to live and to be active. The degree to which free oxygen exists in soil plays a major role in regulating the favorable or unfavorable conditions under which humus is formed. The same is true for the amount of moisture, for the soil temperature, and for the carbon to nitrogen ratio of the residues being decomposed.

Where little or no free oxygen exists (e.g. in stagnant water), decomposition of organic matter occurs by anaerobic organisms.

This process is slower than that conducted by aerobic organisms but can, in the long run, produce a greater amount of humus (e.g. muck or organic soil). Humus formed under water is slightly different than its aerobic counterpart due more to the nature of the residues from the two different environments than the process of aerobic vs. anaerobic humification. Most of the contributions of organic matter to organic soils are from water dwelling insects and microbes that have a higher percentage of protein than the plant residues found in forests, fields, or gardens. Other components come from organic residues transported by wind and water currents to a location where they can accumulate and settle. Also, much of this translocated material may already be humus. Higher percentages of humus are found in soils formed anaerobically because conditions are more favorable for humus accumulation and less favorable for its destruction.

At the other extreme is an environment where there is too much oxygen. If moisture and soil temperature are also at optimum levels, organic matter can be decomposed so quickly that no accumulation of humus will occur at all (e.g. in tropical environments where high temperatures and moisture levels occur in predominantly sandy soils that naturally contain an abundance of air).

Soil temperature is another important controlling factor in the formation of humus. As the temperature of a soil increases there is a corresponding increase in microbial activity. Soils that exist in warmer regions of the earth tend to have lower average levels of humus than soils in colder areas. Figure 2-1 shows that at a soil temperature of eighty eight degrees F, with adequate aeration, humus can no longer accumulate.

MICROBIAL PROCESSING

During humification of organic matter, microbes dismantle most of the sugars, starches, proteins, cellulose and other carbon compounds to utilize them for their own metabolism. The assimilation of these nutrients from the original residues by microorganisms is the first stage in the process of creating humus. Some of the more easily digested components of the residues end up being used by many different varieties of organisms, and may never actually become humus. However, they provide energy and protein for the life cycles

Influence of
Soil Temperature
on Humus Accumulation

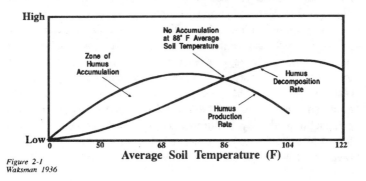

Figure 2-1
Waksman 1936

of the organisms involved in the synthesis of humus. The components of the residues, which are more decay resistant, are not so much assimilated as they are altered by microbial processing into humic substances.

Much of the nutrients and energy assimilated into the bodies of microbes are re-used by other microbes when death occurs. Some is mineralized back into plant food and some is changed into biologically resistant compounds that accumulate as components of humus. As more and more members of the biomass club participate in the festivities of eating, dying and being eaten, the cycles of soil life are implemented. Plants create organic matter that eventually feeds soil organisms, which transform the resources from the residues back into plant food, into nutrients for other organisms and into humus.

The digestion of organic matter in the soil is analogous to the digestive system in animals. Nutrients derived from food ingested by an animal are diffused into its body where they are utilized for energy and production of new cells. By-products, such as urea, water, carbon dioxide and other simple compounds are given off. The indigestible portion of the food is excreted as feces.

In the soil, organic matter is assimilated by microorganisms utilizing the nutrients and energy for their own metabolism. Their activities convert much of the organically bound nutrients back into a mineral form which is usable by plants and other microbes. The indigestible portion of the residues accumulate as humus. However, humus is not completely immune to decomposition. Microbes will eventually recycle all the elements in humus back to where they initially came from, even if it takes a millennium to do it.

DECAY RESISTANCE

Some of the components in organic residues are much more resistant to decay than others. Carbohydrates, such as sugars and starches, will decompose faster than other carbohydrates, such as cellulose and hemicellulose. Fats, waxes and lignins are the most resistant to decay of all the organic components. Proteins vary in decay resistance but are generally more resistant than sugars and starches but more easily decomposed than all the other components.

Average Consistency
of Humus in Mineral Soils

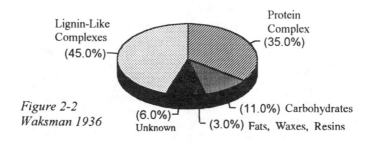

Lignin-Like Complexes (45.0%)

Protein Complex (35.0%)

Figure 2-2
Waksman 1936

(6.0%) Unknown

(11.0%) Carbohydrates
(3.0%) Fats, Waxes, Resins

Although many of these components exist in humus in a biologically altered form (see figure 2-2), the degree to which they exist in the organic residues plays a role in the accumulation of humus. Materials that contain high percentages of easily decomposed components such as sugars, starches and proteins are, for the most part, assimilated back into the living biomass. Although the energy and protein provided by these residues help in the creation of humus, the ratio of their mass and weight to the measure of humus produced is relatively high (i.e. only a small amount of humus is created).

Materials that contain a large percentage of lignins, cellulose or other biologically resistant components have less to offer plants in the way of recyclable nutrients but contribute significantly more to the formation of humus.

Different plants inherently have different ratios of these organic components, but variance also appears in the same plants at different stages of their lives. Green leaves from deciduous trees, for example, have a very different analysis of proteins vs other components than their dry, fallen counterparts. Figure 2-3 shows the changes that occur in rye plants from early growth to maturity. At the young, succulent stage, organic matter from this source would not contribute very much substance for the accumulation of humus but it would benefit more the immediate needs of microorganisms and plants. Near the end of its life, however, the rye plant would add little to the nutrient needs of plants and soil life but provide more raw materials needed for the formation of humus.

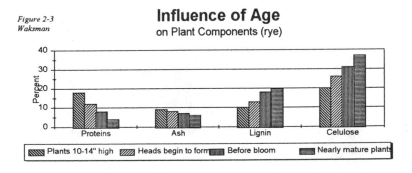

Figure 2-3
Waksman

Influence of Age
on Plant Components (rye)

Biologically resistant components such as lignins, fats and waxes are structurally and chemically changed by microbial processing. Other biologically resistant carbon compounds are created by microorganisms as by-products of their decay activities. These decay resistant compounds are what humus is made of. This is not to say that humus is immune from further decay, but its resistance to decomposition is at a level that enables it to exist for decades, if not centuries, as a soil conditioner, a habitat for microbial life, and a vast reservoir of plant and microbial nutrients.

ENERGY

Humus is essentially a massive storage battery containing energy that was originally derived from the sun. Researchers in England calculated that an acre (furrow slice) of soil with four percent organic matter (stable soil humus) contains as much energy as twenty to twenty five tons of anthracite coal. Another researcher in Maine equated the energy in the same amount of organic matter to 4,000 gallons of number two fuel oil. This organic energy, stored as carbon compounds, was originally derived from the sun by autotrophic organisms such as plants that can extract carbon from atmospheric carbon dioxide. About one percent of the energy from the sun that reaches plant leaves is used to photosynthesize carbon compounds. During the plant's life much of the energy that is absorbed from the sun is utilized for growth, foliage production, flowering, seed production and other functions. About ten percent of the absorbed energy, initially from the sun, is left available to a consumer (e.g. an animal that eats and digests the plant). This leftover energy is called *net primary production*. Like the plant, the animal uses most of the energy it consumes for functions such as growth and sustenance and offers about ten percent of the energy it derived from plants to the next consumer in the food chain.

Subsequent digestions through the food chain continue the rapid depletion of available energy from one trophic level to the next. The final consumers of this energy reside in the soil. In figure 2-4 an

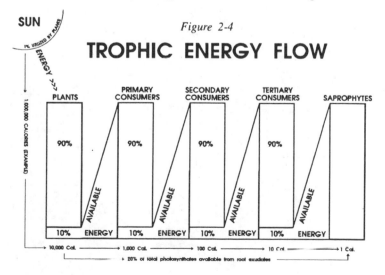

Figure 2-4

TROPHIC ENERGY FLOW

arbitrary quantity of energy has been used as an example to show its flow and use. In this case, the one million calories of energy offered by the sun is reduced to one calorie of available energy by the time it flows through the food chain to soil saprophytes (decay organisms). However, during the season when plants are active about twenty percent of the carbon that they absorb from the atmosphere is exuded through the roots as photosynthesized carbon compounds. They are then utilized by organisms living on, near or within the root surface. This phenomenon provides a direct and constant flow of plant synthesized energy for many soil microorganisms.

Obviously, the level of energy available from plant residues is higher than what can be offered by the remains of herbivores, which is higher that what is available from the residues of carnivores. The various energy levels of different residues stimulate populations of different soil organisms that perform different functions in the soil. Their populations are controlled by the amount and type of residues introduced into the soil which, in turn, controls the quantity and characteristics of humus.

CARBON CYCLE

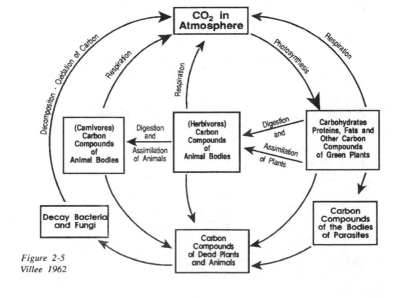

Figure 2-5
Villee 1962

CARBON CYCLE

Throughout this digestion and assimilation process, from the consumption of the sun's energy by the plant to the decomposition of all residues in the soil, carbon is released back into the atmosphere as carbon dioxide (CO_2) through the process of respiration. The evolution of carbon dioxide from organic matter is an integral part of the life cycle. Figure 2-5 examines the cycle of carbon from the atmosphere, through the food chain and back into the atmosphere. Plants and other autotrophic organisms (producers) need carbon dioxide in the atmosphere to live. Without it, no heterotrophic organisms (consumers), which depend on the producers for energy, could exist either. If CO_2 were not evolved, atmospheric carbon would not be available to plants. The accumulation of humus would bury the planet and life, as we know it, could not exist.

Over a one year period and under average conditions, about 60% to 70% of the carbon in fresh organic residues is recycled back to the atmosphere as carbon dioxide. Five percent to ten percent is assimilated into the biomass and the rest resides in new humus. Note: New humus is not necessarily stable humus. It can take decades for humus to develop the biological resistances necessary to be considered stable.

COLLOIDAL PROPERTIES

Colloidal refers to the attraction certain soil particles have for ions of soil nutrients (see chapter 7). Whether organic (humus) or mineral (clay), the colloid is a very small soil particle, often referred to as a *micelle* (meaning micro-cell), and carries a negative electro-magnetic charge that can hold positively charged ion nutrients (*cations*) in a manner that allows plant roots access to them. This phenomenon is called *cation exchange* (see chapter 7).

When decomposed organic matter reaches a certain level of maturity and can be referred to as humus it gains colloidal properties which react, in terms of cation exchange, almost identically to mineral colloids. However, humus can have a far greater capacity to adsorb cations than clay, especially in a soil with a near neutral pH (see chapter 7).

In the soil scientists' quest to isolate and define humus, many terms such as humic acid, fulvic acid, humates, humins and ulmins were developed to help literalize their findings. Some of these terms have relatively complicated definitions and are used to identify the different compounds produced by various chemical extraction methods. Most are too general to define the complexities of humus. However, it is helpful to be familiar with two of these terms, i.e. *humates* and *humic acid*, to understand the colloidal properties of humus.

Average Elemental Analysis of Humus

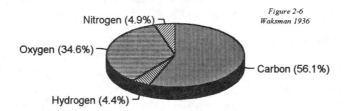

Nitrogen (4.9%)

Oxygen (34.6%)

Carbon (56.1%)

Hydrogen (4.4%)

Figure 2-6
Waksman 1936

When humus particles are formed the chemical composition is predominately carbon, hydrogen, nitrogen and oxygen (see figure 2-6). The hydrogen ions that reside in compounds on the surface of the micelle can be displaced by other cations such as calcium, magnesium, or potassium (see figure 2-7). By chemical definition, any compound that contains displaceable hydrogen is an acid...hence *humic acid*.

If the hydrogen ions are displaced with base cations such as potassium, calcium or magnesium the new compound is considered chemically to be the salt of humic acid or *humate*.

By weight, hydrogen accounts for only a small percentage of humic acid, but because the atoms of this element are the smallest and lightest of all elements its numbers are higher than any other. Each location of a hydrogen ion on the surface of the micelle can potentially become an exchange site for a base cation. The tremendous amount of surface area of humic particles coupled with the high

ORGANIC COLLOID

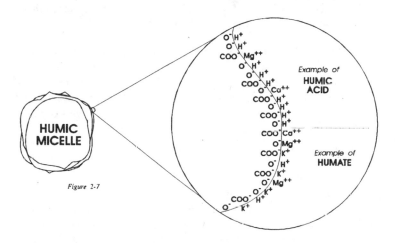

Example of
HUMIC ACID

Example of
HUMATE

HUMIC MICELLE

Figure 2-7

number of exchangeable hydrogen ions (H⁺) can significantly increase the cation exchange capacity (CEC) of any given soil.

In a soil where hydrogen ion activity is high (i.e. low pH), there are usually too few base cations such as calcium or potassium to displace H^+ from humic acid. Humic acids have some ability to react with mineral particles in the soil, liberating base ions such as potassium, magnesium and calcium. As more and more bases (released by chemical and biological activity) exchange with H^+ on the humic colloid, humic acid is chemically changed into humates.

Humates, are essentially organic colloids that are saturated with adsorbed base cations. A rich soil with a near neutral pH might contain a high level of humates. Whereas the same soil with a low pH could be replete with humic acids. Unfortunately, many humic substances can dissolve and leach to lower soil horizons, causing a low pH environment such as in a New England forest where annual precipitation is relatively high.

The presence of stable organic matter in the soil significantly improves the nutrient exchange system by increasing the quantity of organic colloids. These colloids increase the soil's capacity to hold many nutrients necessary for plant growth.

SOIL CONDITIONING

Humus is an amazing soil conditioner. Only three to five percent humus will transform lifeless sand into a rich loam. It has abilities to both bind sand and granulate clay.

In sandy soils, plant and microbial mucilages from humus reduce the size of the pores between sand particles, increasing the moisture holding capacity of the soil and reducing the leaching of soil solution and all the dissolved nutrients it carries. As the moisture content increases, more plants and microbes can inhabit the environment. This, in turn, accelerates the creation of more humus. Under ideal conditions, the advancement of humus in sand eventually will develop the most preferred type of loam for plant production. Unfortunately, conditions for the development of humus in sand are not always ideal. In tropical environments, for example, where moisture and temperature are optimum for populations of decomposition bacteria, organic matter is quickly assimilated back into the biomass. Coupled with the abundance of oxygen in a porous sand, it is difficult, if not impossible, for humus to accumulate.

In clay soils, humus forms an alliance with clay particles. Complexes are formed between the two particles because both particles are colloidal (i.e. they have an electro-negative charge capable of attracting and holding cation nutrients). These complexes not only increase the soil's overall CEC but also mitigate the cohesive nature of clay, by causing granulation.

The accumulation of humus is naturally easier in clay soil than in sand because the environmental conditions for decay bacteria in sandy soils are often not as ideal. Moisture levels in clay soils will often reach the saturation point, leaving little room for oxygen needed by aerobic life. Soil water also acts as a buffer for temperature changes, which keeps the much needed heat level for microbial activity at a minimum. In addition, the evaporation of moisture from the surface actually has a cooling effect on the soil (just as evaporation of perspiration from the skin cools the body). Clay can also assist in the stabilization of humus. The clay-humus complexes formed in the soil can further inhibit bacterial decomposition and can increase the lifespan of humus to over a thousand years. Soil scientists calculate that in Allophanic soils (a volcanic clay soil) the mean residence time of humus ranges from 2,000 to

5,000 years. Over time and under the best of conditions, humus can eventually change both clay and sandy soils into media that are visibly similar.

ACCUMULATION/DESTRUCTION

We can see from the discussion about the carbon cycle in this chapter that it is necessary for humus to, not only accumlate, but also be destroyed so that carbon dioxide is returned to the atmosphere where plants can access it. In an uncultivated, natural environment, humus accumulates in accordance with the favorable or unfavorable conditions of the region. Unless global or regional conditions change, the level of humus accumulation reaches an equilibrium with the factors that destroy it. The humus then becomes a relatively fixed component of that environment.

In cultivated environments, humus is an important asset which, like most other assets, is easier to maintain than it is to replace. Unfortunately, the value of humus is, oftentimes, not fully realized until it is severely depleted and its benefits are no longer available. The trick is to understand the value of organic matter before it is gone.

Old, stable humus, is biologically resistant. Depending upon the environmental conditions under which it exists, humus can sit in the soil for centuries, even millennia, with only a minimal amount of decomposition occurring. However slight, decay still occurs, and eventually even old humus will cycle back from where it came. The formation of new humus is critical to maintaining a stable presence of this asset in the soil.

Figure 2-8 shows a typical response of organic matter introduced into the soil. It is important to note that even under the best of conditions, a relatively small amount of humus is created in comparison to the level of organic matter initially introduced. If conditions exist that further accelerate the decomposition of organic matter, even less humus will eventually be created. In extremes, such as tropical environments where moisture, heat and soil oxygen are abundant, a great amount of carbon dioxide is evolved but not much in the way of humus.

Aeration from the plow or rototiller is probably the most

HUMIFICATION

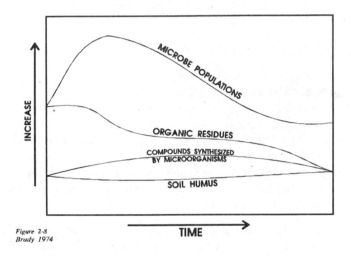

Figure 2-8
Brady 1974

TIME

significant factor in the depletion of native humus levels in cultivated topsoil (see chapter 5). That, coupled with mono-cultural practices and the absence of organic carbon in fertilizer materials, has cause a greater than fifty percent decline in native humus levels over the years on many of the farms throughout the U.S. This represents a loss that probably will never be recovered. Even old humus complexes that are normally very resistant to decay can be fractured by cultivation and made more vulnerable to biological processes.

Excessive applications of lime can significantly accelerate the decomposition of humus (see chapter 5). The low pH in acid soils inhibits the activities of bacteria. However, when lime is applied the pH of the soil rises and bacteria populations grow. Those larger populations cause a relative increase in the decomposition rate of humus. Experiments conducted in 1920 show a marked increase in CO_2 evolution (a measure of organic matter decomposition) as varying amounts of lime were added to soil (See figure 2-9). A small amount of calcium can stimulate plant growth to the point where the increased amount of residues added to the soil balance the loss from greater microbial activity. However, excessive lime applications can hasten the destruction of humus at a pace greater than the plant residues can accumulate it.

Influence of Lime
on Organic Matter Decomposition

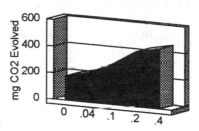

Figure 2 9
Waksman 1936

Excess nitrogen applied to crops is another culprit that can stimulate the activities of decomposition bacteria (see chapter 5). The effect of adding too much nitrogen to the soil is similar to what happens when it is added to a pile of slowly composting carbonaceous organic matter such as dry leaves or saw dust. The temperature of the pile is raised immediately, large volumes of carbon dioxide are released, and the whole process of decomposition is accelerated exponentially. This reaction occurs regardless of the type of nitrogen added (i.e. organic or inorganic).

Fortunately, humus is a renewable resource. Its presence in the soil can be maintained indefinitely. Unfortunately, many agricultural and horticultural practices are essentially mining humus. Like other mined products, such as coal, minerals and oil, the natural resource can, eventually be exhausted.

Humus in the soil has more real value than money, real estate, stocks or bonds. Its value doesn't fluctuate, it doesn't become scarce in a recession, its worth can't be depleted by inflation, and it can't be stolen. It is the direct or indirect source of sustenance for all life on earth. It can sometimes be lost by environmental changes, but more often, its demise results from either the apathy or the inadvertent errors of the steward who tends it.

SUMMARY

It is not realistic to think that one can quantify or qualify humus production from contributions of organic matter to soil. There are too many factors that control its formation and existence in the soil environment. One can only assume that cultural practices that both minimize the depletion and contribute to the formation of humus will maintain the best possible level of soil humus for each environment.

Sources:

Albrecht, W.A. 1938, Loss of Organic matter and its restoration. U.S. Dept. of Agriculture Yearbook 1938, pp347-376

Arshad, M.A. and Coen, G.M. 1992, Characterization of soil quality: Physical and chemical criteria. American Journal of Alternative Agriculture v7 #1 and 2, 1992 pp 25-31. Institute for Alternative Agriculture, Greenbelt, MD

ASA# 47. 1979, Microbial - Plant Interactions. American Society of Agronomy. Madison, WI

Bear, F.E. 1924, Soils and Fertilizers. John Wiley and Sons, Inc. New York, NY

Brady, N.C. 1974, The Nature and Properties of soils. MacMillan Publishing Co. Inc. New York, NY

Brown, B. and Morgan, L. 1990, The Miracle Planet. W. H. Smith Publishers, Inc. New York, NY

Buchanan, M. and S.R. Gliessman 1991, How Compost Fertilization Affects Soil Nitrogen and Crop Yield. Biocycle, Dec. 1991. J.G. Press Emmaus, PA

Gershuny, G. and Smillie, J. 1986, The Soul of Soil: 3rd Edition. Ag Access. Davis, CA

Hillel, D.J., 1991, Out of the Earth: Civilization and the Life of the Soil. The Free Press. New York, NY

Holland, E.A. and Coleman, D.C. 1987. Litter Placement Effects on Microbial and Organic Matter Dynamics in an Agroecosystem. Ecology v68 (2), 1987: 425-433

Huang, P.M. and M. Schnitzer 1986, Interactions of Soil Minerals with Natural Organics and Microbes. Soil Science Society of America, Inc. Madison, WI

Jenny, H. 1941, Factors of Soil Formation. McGraw - Hill Book Co. New York, NY

Lucas, R.E. and Vitosh, M.L. 1978, Soil Organic Matter Dynamics. Michigan State Univ. Research Report 32.91, Nov 1978. East

Lansing, MI

Makarov, I.B. 1986, Seasonal Dynamics of Soil Humus Content. Moscow University Soil Science Bulletin, v41 #3: 19-26

Nosko, B.S. 1987, Change in the Humus for a Typical Chernozem caused by fertilization. Soviet Soil Science, 1987 v19 July/August p67-74

Novak, B. 1984, The Role of Soil Organisms in Humus Synthesis and Decomposition. Soil Biology and Conservation of the Biosphere. pp 319-332

Parnes, R. Fertile Soil: A growers Guide to Organic & Inorganic Fertilizers. Ag Access, Davis, CA

Powlson, D.S. and Brooks, P.C., 1987, Measurement of Soil Microbial Biomass Provides an Early Indication Of Changes in Total Soil Organic Matter due to Straw Incorporation. Soil Biol. Biochem. 1987 v19 (2): 159-164.

Sagan, D. and Margolis, L. 1988, Garden of Microbial delights: A Practicle Guide to the Subvisable World. Harcourt Grace Jovanovich, Publishers. Boston, MA

Senn, T.L. and Kingman, A.R. 1973, A Review of Humus and Humis Acids. Clemson Univ. Research Series #145, March 1, 1973. Clemson, SC

Silkina, N.P. 1987, Effects of High Nitrogen Fertilizer Concentrations on Transformation of Soil Organic Matter. University of Moscow Soil Science Bulletin 1987, v42 (4): 41-46.

Singh, C.P. 1987, Preparation of High Grade Compost by an Enrichment Technique. I. Effect of Enrichment on Organic Mater Decomposition. Biological Agriculture and Horticulture 1987, vol 5 pp 41-49

Smith, G.E. 1942, Sanborn Field: Fifty Years of Field Experiments with Crop Rotations, Manures and Fertilizers. University of Missouri Bulletin #458. Columbia, MO

SSSA# 19. 1987, Soil Fertility and Organic Matter as Critical Components of Production Systems. Soil Science Society of

America, Inc. Madison, WI

Stork, N.E. and Eggleton, P. 1992, Invertebrates as determinants and indicators of soil quality. American Journal of Alternative Agriculture v7 #1 and 2, 1992 pp 38-47. Institute for Alternative Agriculture, Greenbelt, MD

Veen, A. van and Kuikman, P.J. 1990. Soil structural aspects of decomposition of organic matter by micro-organisms. Biogeochemistry, Dec. 1990, v11 (3): 213-233

Villee, C.A. 1962. Biology. W. B. Saunders Company. Philadelphia, PA

Visser, S. and Parkinson, D. 1992, soil biological criteria as indicators of soil quality: Soil microorganisms. American Journal of Alternative Agriculture v7 #1 and 2, 1992 pp 33-37. Institute for Alternative Agriculture, Greenbelt, MD

Wallace, A., Wallace, G.A. and Jong, W.C. 1990. Soil Organic Matter and the Global Carbon Cycle. Journal of Plant Nutrition 1990 v13 (3/4): 459-456

Waksman, S.A. 1936, Humus. Williams and Wilkins, Inc. Baltimore, MD

Waksman, S.A. and Woodruff, H.B., The occurrence of bacteriostatic and bactericidal substances in the soil. Soil Science v53 pp223-239.

Chapter 3

WATER

Water is the fluid of life. It flows through the veins, arteries and capillaries of the planet, carrying the nutrient needs of the biomass. Because water is nature's most abundant compound, its importance is often overlooked. However, when one considers that most living things are made up of approximately 80-90% water, its essential role in our ecology is put into clearer perspective.

Water is a dynamic component of the natural soil system. Not only is it necessary for the existence of all organisms, including plants and microbes, but it also functions as: 1. a solvent, dissolving and mobilizing nutrients in the soil, 2. a regulator, controlling the amount of atmosphere and the temperature of the soil, and 3. a source of hydrogen and oxygen for plant use and soil chemistry. In contrast to its extremely crucial nature is water's incredibly destructive force when nature delivers an overdose in the form of torrential rain or floods.

POLARITY

It is common knowledge that the water molecule is made up of two atoms of hydrogen (H) and one atom of oxygen (O), and is expressed chemically as H_2O. The way in which these atoms are arranged (see figure 3-1) gives water molecules a polarity that is responsible for adhesion (i.e. its attraction to solid surfaces) and cohesion (i.e. its attraction to itself). An example of adhesion is water droplets that cling to a smooth surface such as a window. The strength of both cohesion and adhesion is well illustrated when two panes of glass are held together by a film of water. Attempts to separate the glass without sliding them apart will often result in a broken pane. Water's dynamic relationships in the soil are derived from this polarity.

WATER POLARITY

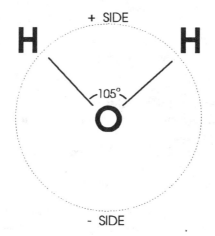

Figure 3-1
Brady 1974

ENERGY

Water has energy referred to as free energy. The concept of energy in water may be confusing because it is not the same kind of energy that exists in fuels, but there are familiar examples that make it easier to understand. A good example of water movement is an experiment with a small spill and a dry paper towel. When the paper towel comes into contact with the water it is instantly absorbed. The movement of the water into the paper towel illustrates the energy difference between the spill and the towel, and the tendency for an energy equilibrium to be established.

It is important to understand that the movement of any substance on earth requires energy in one form or another. Ice cubes melting in a glass of water shows thermal energy flowing from the water to the ice (i.e. the warmer to the colder material). This example demonstrates temperature as the energy that is causing movement but, again, illustrates the natural tendency for an energy equilibrium to be found. Other examples of energy governing the movement of matter are abundant in nature but go largely unnoticed by the busy beings that inhabit the environment.

The energy in water fluctuates in intensity as it reacts to variables in the soil such as plant utilization, the dissolving of nutrients, or evaporation from the surface. Where there is more water in the soil, there is more energy. Physical law dictates that levels of high energy will flow to areas with low energy. So, just as heat flows to cold, the moisture from a wet area in the soil will flow to dryer areas, as long as the soil texture remains relatively consistent.

As water moves from high to low energy areas its cohesive nature drags more water with it and a flow is created. On a hot sunny day when high rates of evaporation are occurring at the surface of the soil, an area of low energy is created which attracts water from beneath the surface through capillaries (i.e. passages in between soil particles). This movement of water is similar to how liquids are drawn up a wick. During a rain storm when the surface contains more moisture than the soil beneath it, the direction of the water movement is reversed.

Precipitation (or lack of) is not the only factor which effects the movement of water in the soil. The existence of plant life contributes as well. As root hairs absorb water, the moisture level in the area immediately surrounding the root is reduced, creating a low energy area. This activity, in turn, attracts new water into the root zone.

The negative energy force that moves water is called suction (so that soil scientists can refer to it in positive terms). The term *capillary action* is also used to describe suction. As the soil dries out, the suction in the desiccated area is increased and water from surrounding areas is drawn to it.

Heavy rains may saturate soils to the point where all the pores and capillaries in the soil are filled with water. In this condition, suction is reduced to zero and gravity becomes the dominant force controlling water movement. Water that is drawn out of the soil by gravity eventually finds its way to rivers, lakes and oceans where it can be evaporated and recycled to the soil in the form of precipitation. After a short period of time (depending upon soil texture), gravity will draw off enough water to create an energy equilibrium between it and the force of suction. This equilibrium also forms a balance of air and moisture in the soil. The moisture level of the soil in this condition is said to be at field capacity.

WILTING POINT

As plants and surface evaporation draw more water from the soil, suction is increased and water is drawn from reserves beneath the surface. If moisture removal continues, but little reserve water is available, the soil will eventually reach a moisture level known as the wilting point; when plant tops can become permanently wilted. The free energy of the soil water at this point is very low which creates very high suction. In figure 3-2, we see the relationship between soil water and suction. As the thickness of the water film attached to soil particles increases, suction decreases and vice versa. At the wettest extreme (*saturation*), there is no suction and other forces such as gravity can dominate the movement of water. At the other extreme (*unavailable water*), suction is high enough to draw moisture from most other sources around it.

WATER DYNAMICS

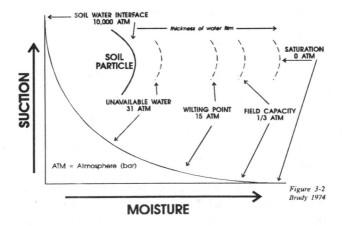

Figure 3-2
Brady 1974

As the soil dries further, the remaining moisture is held tightly by soil colloids (i.e. clay or humus) and other small soil particles (by adhesion) and is unavailable to plants or other organisms (see chapter 7). This phenomenon seems to reserve the last remnants of water for re-wetting purposes. Because of water's attraction to itself (cohesion), any left over moisture will allow the soil to re-absorb water easily.

AVAILABLE WATER

Water that adheres to soil particles attracts other water. As the thickness of the water film adhering to the soil particle increases, the more available it is to plants and other organisms (see figure 3-2). The space between soil particles (pore space) is where available water resides. The amount of water held by soil particles is shown in relation to *atmospheres* (ATM), a measurement of suction, instead of percent saturation because different soils have varying capacities.

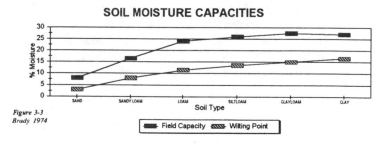

Figure 3-3
Brady 1974

The field capacity and wilting point of a soil is determined by: 1. the amount of pore space, 2. the amount of colloids, and 3. the average particle size, that is the soil texture (see figure 3-3). In a clay soil, porosity is at a minimum, but the level of soil colloids is very high, and the soil particle sizes are very small (i.e. large amount of surface area). Although the attraction for water is great, the availability of that water is much lower than in soils with more pore space. Humus has a similar attraction for water, and like clay, can hold it tightly. However, the soil granulation caused by organic matter creates pore space that can hold greater quantities of available water. At the other extreme is sand which has a tremendous amount of porosity but, because of the low level of colloids and the large particle sizes (i.e. less surface area), it is unable to hold a significant capacity. Under these circumstances gravity has a distinct advantage in its tug-of-war against suction. At the other extreme are soils with very small particles (e.g. clay). These soils have tremendous surface area and extremely small pore spaces. Not only is it difficult for gravity to drain excess moisture from these soils, but plants will often have a hard time getting the water they need.

SOIL HORIZONS

All of this information is enlightening and useful for understanding soils that remain consistent to depths of a few feet or more, unfortunately, there are very few places in New England where that occurs. Water movement is significantly disrupted when it meets with a soil horizon that has a different texture. An example of this phenomenon is an experiment that was done many years ago where several inches of loam was placed on top of a thick layer of sand in a glass case. As water was slowly added to the surface of the loam, it was evident through the glass that the water would prefer to move horizontally through the loam than down into the sand. The loam would reach near saturation before gravity could pull any water down into the sand. This is because the force of suction is greater than gravity until the saturation level is reached. This reaction would occur with a number of different textured materials beneath the loam.

The research suggests that abrupt changes in soil consistencies from one horizon to another can affect both the movement of water and the soil's ability to store it. Thus, thin applications of topsoil to a well drained base may do more harm than good in terms of water dynamics.

WATER MOVEMENT THROUGH PLANTS AND ATMOSPHERE

The movement of water in plants is governed by the same physical laws as it is in the soil. The plant does not have a heart that pumps body fluids about. The moisture level in the soil must be greater than in the plant for suction to occur through the roots. The same rules apply to the stems, to the branches, and to the leaves. Water that is evaporated from the leaves of plants to the atmosphere reduces the energy level in the leaves, which then creates the suction of water from the stems and/or branches. When moisture is drawn up into the leaves there is a reduction of energy in the stems that draws water from the roots which, in turn, draws it from the soil. In all cases, there is more energy in the source than in the location that the moisture is moving to. As energy levels equalize in the different locations, the force of suction diminishes.

The last link in the moisture cycle (see figure 3-4) is the atmosphere, which is also regulated by the concept of free energy. If moisture levels in the atmosphere are low, such as on a warm, dry, windy day, the movement of water through the cycle is accelerated. Conversely, on still days with high humidity, water evaporates from the soil and plant leaves more slowly. The difference between the moisture level in the soil and that in the atmosphere is called the *vapor pressure gradient,* and this controls the rate of water movement through the cycle. On a cool, foggy morning, the energy levels in the soil, plant and atmosphere may be close to equal. Because of the balanced energy levels the movement of water through the cycle is slower and, at times even stopped. One of the new technologies in moisture control systems for greenhouses is a fog generator which creates an atmosphere with 100% relative humidity. In this environment the transpiration of water through the plant is slowed and often stopped. Because of the high energy levels in the

WATER CYCLE

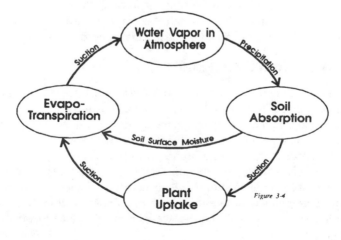

Figure 3-4

atmosphere, the water moves from the air to the soil instead of in the normal direction caused by evaporation.

The leaves of plants provide further regulation of the cycle. Leaves contain small pores called stomata that can open and close to regulate the amount of water and gases that flows through them. On hot, dry days, when evapo-transpiration is highest, the stomata can contract, slowing down the flow of moisture through the plant.

OSMOTIC SUCTION

Nutrients that dissolve in soil water utilize some of its free energy, which can inhibit its flow through the moisture cycle. This phenomenon is referred to as *osmotic suction,* and is demonstrated by water's migration to soluble salts. For example, rural municipalities will often apply calcium chloride (a soluble salt) as a dust control on gravel roads. Moisture is drawn from the humidity in the air to the dry salt which dissolves and seeps into the surface of the road. The osmotic suction created by the concentrated solution continues to attract moisture from the air because its energy level is lower than the water vapor in the atmosphere. The reversed flow of moisture from the air to the road continues until the salt is diluted enough to reduce the force of osmotic suction. This reversed flow of moisture into the road provides temporary dust control.

In the soil, water that is rich in dissolved nutrients has less energy and moves more slowly into low energy areas such as plant roots. If the concentration of dissolved salts becomes too strong, the movement of water into the plant may be less than is needed to sustain life.

WATER VAPOR MOVEMENT

Relative humidity refers to atmospheric moisture relative to air temperature. The warmer the air, the more moisture it will hold. As water vapor in the atmosphere is moved upward, coincidentally by the same energy forces that move it through the soil and plants, it condensates in the cooler altitudes to form clouds. Condensation creates heat which forces the moisture to rise into higher altitudes with even cooler temperatures where more water vapor will condensate. Precipitation forms as the clouds reach moisture levels that can no longer defy gravity, and the water returns to the earth where it begins its journey again.

EROSION

Frequently, when it rains, water is applied to the soil at a faster rate than it can be absorbed. The excess can either accumulate and puddle, or it can escape as surface runoff depending on the topography. In a sloping environment, the movement of water over the surface of the soil can remove and relocate a tremendous amount of mineral and organic material. Under mild erosion, the material may be deposited only a few feet away. However, in severe conditions soil may be carried all the way to the ocean. The amount of soil removed per year from erosion can be as high as forty tons per acre. In addition to the loss of soil, the nutrients that are removed will often far exceed what an annual crop would extract from that soil.

Nature provides its best protection from erosion with plants and organic matter. The roots of plants adhere to soil particles and prevent much of the erosion. Plant tops can also intercept a significant amount of rain water where it evaporates before reaching the soil. It is estimated that one third to one half of the precipitation in a forest environment is intercepted and evaporated without ever reaching the ground. Organic matter, because of its adhesive qualities, can hold soil together and significantly reduce the amount of soil removed by erosion (see figure 3-5). The more organic matter in the soil, the more protection from erosion.

General Influence of Organic Matter on Erosion

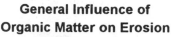
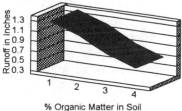

Figure 3-5
Jenny 1941

% Organic Matter in Soil

CONTROLLING SOIL MOISTURE

There is nothing anyone can do to control climate, which is the source of moisture for the soil. However, there are cultural practices that can maximize or minimize the water that is available.

In soils and climates where maintaining an adequate moisture level is a problem, cultivation techniques must be examined to determine where the moisture cycle can be decelerated. Does the variety of cultivar have an inordinately high demand for water? Is there bare soil exposed to the sun? Are there adequate organic matter levels in the soil? Do management practices include excessive tillage or ridge tillage that creates additionally exposed soil surface area? Is the topography appropriate for the crop?

The answers to these questions may give clues as to where the movement of water can be best controlled. They may also indicate an inappropriate location for the current cultivation. In many of the arid agricultural regions of the world, plants are grown under irrigation where they could not naturally exist. Understanding the dynamics of water can help maximize the efficiency of water usage. In lawns, for example, something as simple as raising the height of the mower could shade the soil enough to make the difference between a green lawn and a brown, dormant one.

In conditions where the soil retains too much water and not enough air because of texture and/or topography, the same questions asked above may provide clues to obtain the opposite objective (to accelerated the flow of water through its cycle). Again, the answers may indicate an inappropriate location for the natural existence of a particular species of plant. Artificial drainage may be the only solution. The individual must determine for himself and his environment the best course of action to take.

SUMMARY

The movement of water, like most other phenomena in nature, is controlled by many different factors that, oftentimes, include un-known variables. All of these factors effect the balance of energy between water and its immediate environment; and water's energy constitutes its ability to move through a system of which it is an integral part. Too often the importance of water on earth is overshadowed by its abundance; however, it is necessary to learn more efficient management techniques before, not after, water becomes a commodity in short supply.

Sources:

Bear, F.E. 1924, Soils and Fertilizers. John Wiley and Sons, Inc. New York, NY

Brady, N.C. 1974, The Nature and Properties of soils. MacMillan Publishing Co. Inc. New York, NY

Huang, P.M. and M. Schnitzer 1986, Interactions of Soil Minerals with Natural Organics and Microbes. Soil Science Society of America, Inc. Madison, WI

Jenny, H. 1941, Factors of Soil Formation. McGraw - Hill Book Co. New York, NY

Ray, P.M. 1972, The Living Plant. Holt, Rinehart and Winston, Inc. N.Y., NY

Chapter 4

THE CONCEPT OF PESTS

"Each time we design a solution to a single problem without consideration of the relationships between that problem and the larger context, we only create new problems."

(Brown et al 1976)

This quote seems to have been written specifically for agriculture and horticulture. In plant management terms, it often means *the more you do, the more you'll have to do.* Unfortunately, the expression presents a problem for pest managers because there really is no way to know the larger context in its entirety, or to predict the eventual ecological reaction to pest control activities.

The utilization of controls for pest problems are, almost always, a treatment of symptoms. Sometimes, as with the common cold or flu, it is the only alternative to suffering. Oftentimes, there is an underlying cause creating the problem and in many cases the answer is quite obvious. Corn growers, for example, understand that their weed problems could be controlled by rotating fields into sod crops for a period of time; however, time and equipment constraints make it easier and less expensive to use herbicide. The same is true for lawn care companies who could control many weeds simply by raising the height of their lawn mower. Unfortunately, their customers want the turf to look as much like a putting green as possible.

Like weeds and plant disease, insects that attack crops should also be recognized as a symptom. Often, they may be a symptom of circumstances that are beyond one's control. For example, they may simply result from the *wrong crop, in the wrong place, at the wrong time*, a situation many growers have experienced and only a few can predict. In many cases, it may be impractical to alter the root cause of the infestation because of time, economical, or equipment constraints.

Contrary to popular belief, nature did not put pest organisms on earth specifically to antagonize the human agrarian. Like all other organisms, the individual purpose of the pest is to survive and reproduce, but they also play a higher role in the natural system. In the process of natural selection, insect, disease organisms and even weeds, to a certain extent, cull out much of the weak or damaged plants so that the strongest of the species can survive and reproduce. In many situations pests serve to protect the natural diversity of an area by controlling some plant's ability to completely dominate the environment. Pests will oftentimes control the occurrence of other pests or pest plants (weeds). Wild grape vines, for example, are often decimated by Japanese beetles (Popillia japonica). Unfortunately, the proliferation of one species of pest can support a relative explosion of another and the latter species may be less desirable to horticulture or agriculture than the former. The important point is that nature has a system of checks and balances and the purpose of the pest is often obscured by the inconvenience it causes to man.

A herbivorous pest is simply an organism that is feeding on the wrong plant. The Monarch Butterfly larvae, for example, would be considered a pest if the milkweed plant were of any value to agriculture or horticulture. Instead, this adult insect is a thing of beauty and pleasure to most people. Pests, in general, are organisms that interfere with human enterprise activities. If we ranked pests simply by how much they disrupt an environment, humans would be at the top of the list.

EQUILIBRIUM

Pest management would be simple if all one had to do was to create a healthy soil. It is not uncommon to see many problems disappear, some major ones, when soil quality is improved. The big advantage to a healthy soil is the diversity of life that inhabits it and its intense competition for resources, which oftentimes include, each other. Many plant pests have natural enemies that will exist in far greater numbers in a healthy soil.

Even in the best of soils and under the best of conditions there may be a lingering pest problem that tempts us back to the bottle of biocide. Unfortunately, the use of controls can often lead to a further disruption in the natural balance between pests and their antagonists, competitors and predators. Research has shown, for example, that the suppression of one group of fungi by pesticides leads to a relative increase in another species. Researchers also found that soil respiration (i.e. the recycling of carbon into the atmosphere, *see chapter 2*), is temporarily inhibited by the application of a broad spectrum fungicide. Atmospheric carbon is a nutrient plants require in great abundance. Inhibiting its natural production may cause stress to plants that increases susceptibility to other problems.

Similar circumstances may occur with insecticides. The application of an insecticide often causes a resurgence of the target pest because of predator suppression, changes in biological tolerance by the pest, biocide induced stress in the plant or a combination of all three. Mites, who are notorious for exploiting stressed plants, will react to many miticides by increasing egg production. The result is usually a larger population of mites feeding on stressed plants in the absence of predators. Often the consequences of chemical intervention are not what was originally intended.

If every organism on earth has a purpose, regardless of its importance, then the eradication of any organism is analogous to removing a part from a machine. Sooner or later the consequence of the missing part will manifest itself, often as a new problem. If the new problem is dealt with by the removal of yet another part, even more problems will be created. If the biological system were viewed as if it were a machine, the conventional approach would only be logical if one was completely aware of the entire system, and could predict that the overall effect of the treatment would be favorable. There just isn't anyone who possesses all this knowledge.

In a natural environment, equilibria are established. Even in the unlikely extreme where a pest appears without coincidence of a natural predator, it will eventually consume its own environment, and disappear from a lack of resources. More commonly, the appearance and proliferation of a particular species of pest usually attracts predators, parasites or pathogens that become prolific from luxury consumption and eventually they dominate the area. For example, an infestation of chinch bugs (Blissus leucopterus)on a lawn will eventually bring about a healthy population of big eyed bugs (Geocoris bullatus) to feed on the pests. Since chinch bugs and big eyed bugs have very similar physiologies, most insecticides that affect the pest will also affect the predator. Once the predator is gone, the environment either becomes chemically dependent on the insecticide or succumbs to the destruction of the pest.

Ignoring the problem is not a prescription for pest control. However, it is important to understand the concept of equilibrium in nature. If the application of a biocide wipes out the incidence of natural predators and stresses the target plants, then the eventual outcome is a more favorable environment for the pest.

The control of plant diseases is another good example of the same principle. Most of the fungicides used to control pathogens are also eliminating beneficial fungi that compete antagonistically with the pest organisms, as well as other beneficial fungi. Examples of this include a family of fungi called *mycorrhizae* that help plants forage soil depths for water and nutrients, and saprophytic fungi that help recycle organic matter. Some soil borne fungi have insecticidal properties and can control many plant damaging insects. Recent research has found that, in many cases, the use of fungicides will ultimately increase the incidence of the pathogen it was meant to control. Apparently, some fungi can sense their own demise and, in an attempt to preserve the species, produce an abnormally high number of spores, that are not normally killed by fungicides before they die. The end result is a bigger problem.

It is important to understand that the vast majority of the different species of organisms in any given environment are neutral or beneficial to horticulture or agriculture. Most pesticides do not discriminate enough to avoid affecting some of those species.

Many managers have discovered that the incidence of disease diminishes significantly when the area is biologically enriched with organic matter such as aged compost. Apparently, the amount and type of organisms living in the compost are very competitive with pathogenic fungi (organisms that cause plant diseases). Others have found that soil management practices that encourage the development of organic matter and soil microorganisms also has a similar effect.

The incidence of organisms in soil is relative to the amount of organic matter (OM) content. OM levels are controlled primarily by plant growth and climate; however, farm cultivation practices such as tillage, liming and the introduction of soluble N fertilizers will significantly accelerate the decomposition process (see chapter 5). Plants are the source of raw materials in the formation of OM and are the main influence for the development of microbial biomass. The roots of plants release approximately 20% of the organic compounds produced by photosynthesis into the soil, creating an area of intense biological activity (see figure 4-1) around the roots. This activity serves many different functions that affect the soil, plant and atmosphere. Antagonism and competition directed at pest organisms are only two of them.

INFLUENCE OF PLANT ROOT
ON BACTERIA POPULATIONS

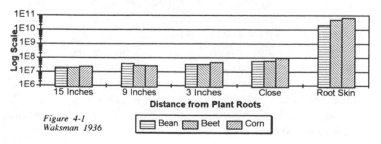

Figure 4-1
Waksman 1936

Perennial monocrops such as fruit trees are generally more susceptible to pest problems because they foster the development of a permanent cycle for the herbivore. In this type of situation, maintaining a competitive balance of organisms is very important. Predators, antagonists and competitors can also become part of the cycle. An orchardist in Vermont has successfully experimented with maintaining an ecological balance for insect control in his apple orchard. The orchard floor is mowed only once per year and the woods surrounding the apple trees have been cut back by a wide margin. Predators abound in the thick and wild environment under the trees. The buffer zone around the orchard protects the trees from plum curculio (Conotrachelus nenuphar) and other migrating pests. Although he is not strictly organic, he currently uses no insecticides on fifty acres of apples. A Canadian researcher found that the parasitism of tent caterpillar (Family Lasiocampidae) and codling moth (Cydia pomonella) increased by 1,800% in orchards with rich floral undergrowth (Leius 1967). Providing a better habitat for predators usually creates a more balanced environment.

Even seemingly innocuous fungicides such as baking soda or elemental sulfur can have a profound effect on the equilibrium of the environment they are applied to. Researchers who have experimented with baking soda (sodium bicarbonate) as a fungicide have had excellent results controlling certain diseases on broccoli and other vegetables. They theorize that baking soda could have a similar effect on almost any fungal disease. However, it is not known exactly how the control works, which worries some researchers who are concerned about nature's balance in a given environment. No matter how innocuous the control may seem to the environment, it can still disrupt the natural system.

In the broad scheme of natural phenomena, the occurrence of herbivorous pests may be an indirect response to the environment's need for more organic matter. When OM levels in a given area are low, the competition offered by beneficial organisms is also low. Infestations of plant feeding insects, of disease, or even of weeds contribute (although unintentionally) large amounts of plant debris to the soil. As OM levels rise, there is a relative increase in the number of organisms capable of either direct or indirect intervention in pest activities.

SOIL FERTILITY

Plants grown in soils with optimum fertility, showing little or no stress associated with nutrient or water shortages, have a significantly lower incidence of (herbivorous) pest damage. The plant's resistance is higher and so is the incidence of predators. The reasons are not clear as to why a predator increase occurs, but trials have shown a notable difference in predator activity where soil fertility is at optimum.

Of all the plant nutrients, excess or insufficient levels of nitrogen (N) and potassium (K) in the plant seem to have the greatest effect on insect activity. Excess nitrogen accumulates as soluble amino acids and nitrates, needed components for the formation of protein by insects. On the other hand, inadequate N in plants arrests growth and slows the movement of protein and sugars through the plant, which causes accumulations of these compounds in leaves and stems. The weakened plant with high concentrations of nutrients must be analogous to herbivore heaven.

Deficiencies or excess levels of potassium have an indirect effect on insect activity due to their relationship with nitrogen. Potassium acts in conjunction with nitrogen to accelerate the processing of soluble amino acids and nitrates into proteins by the plant. The less available nitrogen there is in the plant, the less attractive it seems to be to herbivorous insects and, possibly, to plant pathogens as well.

Although no specific link has been made to other nutrient deficiencies and pest activity, it is reasonable to assume that all components of nutrition for both the plant and soil organisms are equally important. For example, micro-nutrients that are needed in only trace amounts play a major role in the creation of over 5,000 different enzymes necessary for life functions.

The soil's physical condition, which includes structure, density and fertility, can have a big impact on the diversity of soil organisms. Heavy wet soils with little porosity can inhibit the existence of many beneficial aerobic organisms. At the other extreme, sandy soils are often lacking enough moisture to support significant populations of beneficial organisms.

The balance of organisms and fertility by the introduction of beneficial soil components such as compost can be tipped the other way by overdosing. Gardening fanatics who douse their soils with too much of everything, even *good* amendments such as compost, can also create an unnatural environment prone to problems. The organic media introduced may favor one type of organism that, in balance, plays an important role, but in abundance can cause some problems.

Luxury consumption of plant nutrients is rare in nature and can be as unhealthy to plants as it is to people. Available nutrients levels that are either long or short of optimum can cause stress to plants, and lower their resistance to pathogens and to insects. For example, unprocessed nitrates in the vascular system of plants, resulting from excessive applications of soluble nitrogen, is an open invitation for both insects and disease pathogens. Insect bodies are almost fifty percent protein, the synthesis of which demands a good source of nitrogen. Herbivorous insects must be able to locate an abundance of nitrogen to survive. A 1983 study done in India shows seventeen of twenty three plots fertilized with nitrogen resulted in an increase in insect damage. The fine line between too much and too little nourishment is as important to understand for plant health as it is for human health. Unfortunately, common sense is seldom applied to this subject.

Factors that we can not control such as extremes in precipitation, and temperatures also play a major role in the incidence of pest organisms. A crop of peas, for example, may do extremely well in a moderately dry year, and fail miserably from disease if the season is too wet. Obviously the weather cannot be controlled, but the level of tolerance each plant has for extremes in the climate can be increased by producing a strong, healthy plant in a rich, and enlivened soil. Soils with an abundance of organic matter not only help accomplish this but also buffer extremely wet or dry climates by increasing the soil's water holding capacity.

PLANT DEFENSES

Scientists have discovered that healthy plants are not completely defenseless against pest organisms. Most species of plants can synthesize chemical defense compounds that inhibit pest activity. The synthesis of these organic chemicals depends largely on: 1) the availability of the necessary elemental components (i.e. fertility), and 2) the plant's overall health, strength and vitality. In some cases the introduction of a simple primary nutrient such as potassium will cause the host plant to create stronger resistance to certain pests. Plant resistance to pests is manifested in four basic forms. The first is the manufacture of toxins that effect the biology of the pest. The second is its ability to make itself less attractive either by smell, taste or nutritional value, which forces the pest to go elsewhere for sustenance. The third is the development of tolerances to pest damage, i.e. the ability to thrive in spite of the adversity. And, the forth is a physical resistance created by thickening the outer layers of both stems and leaves. Often, a plant may exhibit all four resistances.

It was also discovered that many plants can actually sense imminent insect pest problems and react to them. When some types of plants are attacked, they release a natural plant hormone called ethylene. This compound is often sensed by adjacent plants which react by producing defense compounds in anticipation of an attack. Plant species that have no defenses, either their own or from symbiotic organisms such as predators, would have become extinct long ago.

Researchers have studied thousands of varieties of plants that show some type of control of different pests. Unfortunately, there are more than 20,000 different pests that damage crops and less than 1,000 have been studied, but research continues. It is a relatively safe assumption that other plant varieties with pest control properties exist but remain undiscovered.

Researchers believe that there are as many natural toxins in the environment as there are different organisms. These toxins are usually formed by organisms such as plants, fungi or bacteria, but some may be mineral in nature. Penicillin, for example, is an antibiotic that is toxic to many different types of bacteria, and is made by soil fungi. These toxins are produced as metabolites of biological activity or as compounds formed from the decay of various organic materials. They serve to inhibit survival of a broad range of organisms. Unfortunately, the production and target of these toxins is relatively indiscriminate in terms of crop production. Some species of organisms produce compounds that are toxic to themselves, thus providing their own control for overpopulation.

In some cases, substances that are toxic to one organism may be nutritious to another. The best balance of these natural toxins in the soil is both directly and indirectly related to the level of organic matter (OM). The direct relationship is the toxins produced from the decay of various plant residues. The indirect factor is the relative abundance of toxin producing organisms that live in soils with high levels of OM. Many of nature's toxins have been isolated, synthesized and concentrated by scientists into lethal substances called pesticides. Introducing these substances into the environment in unnatural concentrations often causes an ecological chain reaction with unpredictable consequences.

Plants will display a variety of reactions to grazing by pest organisms. The amount of organic nutrients released into the soil by roots generally increases, creating a relative increase of microbes in the root zone. Microbial activity, in turn, can increase the amount of available mineral nutrient for the plant (see chapter 1). Most plants respond to above ground damage by increasing root growth and the storage of nutrients in the root system. Apparently, reducing the amount of nutrients located in the shoots and leaves makes the plant tops less nourishing and therefore less attractive to grazing insects. The stored nutrients in the root system can increase the plant's ability to recover from pest damage. Unfortunately, this concentration of energy and nutrients in the roots may also attract below ground pests. Chinch bugs (Blissus leucopterus) may, for example, injure the tops of turfgrass plants and, unintentionally, make the roots more palatable to grubs (scarab larvae). Many plants have the ability to significantly increase the production of alkaloids during insect attack. These substances are often toxic or distasteful to the herbivorous pest.

There are also natural organisms called endophytes that are toxic to many foliar feeding insects. These microscopic fungi are found living symbiotically within the cells of certain turfgrass varieties. Seed growers have successfully bred these varieties to make endophytically enhanced turfgrass seed available for general use. High endophyte content in turf is an example of how nature's controls can be utilized for positive horticultural use.

DIVERSITY

Many researchers are experimenting with diversity and the breeding of disease and insect resistant varieties. This approach is close to the method mother nature uses to establish equilibrium. In a wild setting, such as a forest, many species of plants are naturally introduced. The survival of the species depends on its ability to compete, not only with the other plant species, but also with all the environmental factors including: animals, insects, pathogens and climate. If the species survives, its next hurdle is to evolve with the changes in the environment either through natural selection or through adaptation. The development of natural defense compounds or symbiotic relationships with other organisms capable of defense is the basis of this research.

The development and use of both disease and insect resistant varieties of plants is the fastest growing of all the pest control techniques. This science has a long term balancing effect because it provides an overall reduction of nutrient resources for both noxious insects and pathogens. The amount of available food is a limiting factor in the population growth of these pest organisms.

There is an apparent correlation between the lack of diversity and many different pest problems. As cultivations become larger and more crop specific, pest problems appear to grow in intensity. For example, turf with only one cultivar usually has more pest problems than a lawn with many different varieties of grasses. Agricultural fields that are rotated from sod into row crops and back into sod are usually easier to manage, in terms of pests, than the mono-crop operations.

In a natural forest setting, diversity is the key to plant survival and soil conservation. Many plants have pest control properties that can protect themselves and other species. The more diversity in an environment, the better equipped it is to deal with pest problems.

Diversity is also important to consider in the sub-visible world beneath the soil's surface. A healthy and diverse population of micro-organisms, as well as, meso-organisms and macro-organisms is needed to establish a biological equilibrium This diversity prevents any one organism from dominating the environment. The overuse of pesticides can alter that equilibrium to a point where the environment is ultimately much more vulnerable to pest domination.

Adequate amounts of nutrients and micro-nutrients are needed to support a strong and diverse community of soil organisms. The pH of the soil has a profound effect on bacteria populations as well (see figure 4-2). Each of these problems can be remedied with additions and/or cultivation of organic matter.

Effects of pH on Soil Life

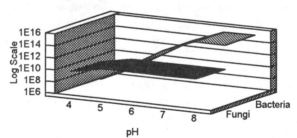

Figure 4-2
Waksman 1936

WEEDS

Weeds are another pest organism that appear as symptoms of different conditions. The natural tendency of plants is to utilize every available space on the soil's surface. If there is sun, water and nutrients in a space with little or no competition for those resources, then eventually, there will be a very happy plant. With the exception of desert, arctic or shoreline environments, it is very rare to see bare ground in a natural setting. A weed is a plant that is in the wrong place. Weeds do not exist in wild settings such as unmanaged forests. Furthermore, many so-called weeds are cultivated and sold as valuable plants for different locations. An expensive installation of a ground cover that creeps into a lawn becomes a weed at the garden's border.

In agriculture, the crop manager objects to weeds because they compete with the cash crop for light, water and nutrients. In horticulture, the same is true to a lesser degree. The greater problem in a landscape is an aesthetical one.

Mother nature does not recognize any plant as a weed. However, she can provide conditions that are more hospitable to certain species of plants, giving them a competitive edge. There are some natural toxins with herbicidal properties that exist in the soil and allelopathic compounds produced by plants. In most cases however, they play a relatively varied and diverse role in controlling the selection of plant species for a given environment. The general lesson nature teaches is to strengthen the preferred species rather that weaken the unwanted.

Soil conditions can have a profound effect on the strength and competitive abilities of certain plants. The weeds that exist in turf, for example, are often a result of a soil conditions such as compaction or excessive porosity (sand) that are incompatible for the preferred turf cultivars. Intervention at the soil level can often provide a remedy that is effective and long lasting. Herbicides may eradicate the weed but they will not change the conditions that caused the problem in the first place. Additionally, many herbicides have been shown to increase pest insect activity by suppressing predators, by stressing plants, or by doing both.

Sometimes, weed control in turf is as simple as raising the height of the lawn mower. This action enables the lawn to be more competitive by utilizing the sun's energy for itself and blocking it from fueling unwanted species of plants. For every eighth of an inch that a lawn mower is raised, there is a thirty percent increase in leaf surface area. That increase causes a relative increase in photosynthesis which feeds a larger and healthier root system. The roots of many turf varieties often develop new plants that further thicken the stand, providing even more competition.

Weeds in agriculture are a different problem. These plants are opportunistic, taking advantage of the available space, water, nutrients and sunlight. Unfortunately, plants do not share these resources equally; they compete for them. In many instances, a cash crop can be completely destroyed by a stronger, undesirable, species of plant. Continuous cropping of a specific variety of plant can create an environment conducive to the establishment of weed cycles. It can also create a soil environment that becomes less suitable for the intended crop each year that it is planted. Often a specific crop grown year after year in the same ground will deplete the soil of, possibly obscure, but critical elements or compounds. This condition in the soil gives weeds without those specific needs a competitive edge.

Crop rotations and cover crops provide the competition needed to suppress noxious weeds in row crops. Mulches, either living or dead, can provide further protection. The University of Vermont recently completed a three year comparison of cultivation vs herbicide in field corn and found cultivation to be more economical. In addition, the absence of herbicide may contribute to a more balanced ecology that protects the corn from other pest problems.

Weeds may also play a greater role in the broad scheme of things. There is strong evidence that the incidence of insect parasites, predators and antagonists exist in far greater numbers where there is an increased diversity of plants, i.e. weeds. The Vermont orchard that uses no insecticide is an example. Many weeds act as bait plants drawing herbivores away from cash crops using various attractants (kairomones). In many cases weeds established in row crops are not competing for a significant amount of resources but are more of an aesthetic problem. Often, the elimination of these weeds creates a more favorable environment for pest insects. Researchers have studied weed integration in crops as a means of insect control and found the practice to be fundamentally possible. However, the habitat and relationships that weeds, insects and crops have are complex and there is not, as yet, a successful formula that applies to all environments.

STRESS

Stress can occur from many different sources other than biocides. The plant inflicts stress upon itself from functions such as flowering, seed development and germination. Other natural forms of stress include drought, heat, cold, heavy rain and wind. A less than optimum soil environment lacking proper structure, fertility, pH or organic matter content can also increase stress. Stress also occurs from air pollution, changes in global climates, acid rain, soil depletion, ground water pollution and other anthropogenic contaminations. Cultural practices such as pruning, transplanting and propagation are sources of stress as well.

Plants, like most other organisms, utilize energy to combat stress. As available energy diminishes, plants can succumb more easily to opportunistic parasites, such as pests, that survive by attempting to secure nutrition while expending as little of their own energy as is necessary. Stressed prey is easier prey. Organisms instinctively know that in order to survive they must absorb more energy from food than they use to find or catch it. Stressed organisms, even humans, are more susceptible to health problems. The less energy a plant spends on stress, the more it will have for growth and defenses; that is to say its basic survival.

Cultivation of any kind is natural only to the human species. The propagation of plants in an environment where they would not naturally grow is an upset in that ecosystem that could eventually lead to pest problems; especially if the introduced plants cannot achieve their ultimate level of strength and resistances because of stress associated with environmental conditions. A plant sprouted in Florida, for example, and shipped to New York State to be planted may use up too much of its own energy reserves coping with introduced stress conditions to be able to survive the natural pests that prey on weakened plants. However, the problem appears to be the pest and not the plant. If biocide is used the apparent problem may disappear, but the ecosystem has been altered. At this point there is no way of knowing how this change will affect the future of cultivation in that environment. By changing the ecosystem, more stress may have been introduced to not only the plants but to beneficial organisms of that environment. The cycle of combating stress by introducing more of it will eventually climax. The end result is difficult to predict, but it is reasonable to assume the environment could, at some point, cease in its suitability for plant production. The development and maintenance of a balanced biomass should be a primary concern of all pest managers.

SUMMARY

In the period from 1945 to 1985, the production of pesticides has increased to over 1.5 billion pounds annually (see figure 4-3). Ironically, damage to crops caused by pests has increased by an average 20% over the same period. The implications of this study should have much more meaning to the land manager than to the environmentalist. It is clear that the use of pesticides is not providing all the control it was intended for.

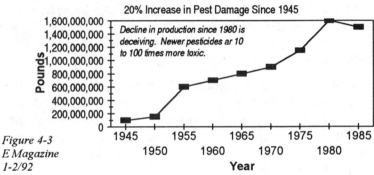

Synthetic Pesticide Production
From 1945 to 1985
20% Increase in Pest Damage Since 1945

Decline in production since 1980 is deceiving. Newer pesticides ar 10 to 100 times more toxic.

Figure 4-3
E Magazine
1-2/92

Sources:

ASA# 47. 1979, Microbial - Plant Interactions. American Society of Agronomy. Madison, WI

Baker, R.R. and Dunn, P.E. (Editors) 1990, New Directions in Biological Control. Alan R. Liss, Inc. New York, NY

Barbosa, P. and Letourneau, D.K. (Editors) 1988, Novel Aspects of Insect - Plant Interactions. John Wiley & Sons, New York, NY

Barbosa, P., Krischik, V.A. and Jones, C.G. (Editors) 1991, Microbial Mediation of Plant - Herbivore Interactions. John Wiley & Sons, Inc. New York, NY

Brady, N.C. 1974, The Nature and Properties of soils. MacMillan Publishing Co. Inc. New York, NY

Brown, B. and Morgan, L. 1990, The Miracle Planet. W. H. Smith Publishers, Inc. New York, NY

Brown, H. Cook, R. and Gabel, M. 1976, Environmental Design Science Primer. Earth Metabolic Design. New Haven, CT

Chet, I. (Editor) 1987, Innovative Approaches to Plant Disease Control. John Wiley & Sons, Inc. New York, NY

Davidson, R.H. and Lyon, W.F. 1987, Insect Pests of Farm, Garden and Orchard: Eighth Edition. John Wiley and Sons, Inc. New York, NY

Ehrlich, P.R. and Ehrlich, A.H. 1981, Extinction: The Causes and Consequences of the Dissappearance of Species. Random House, Inc. New York, NY

Fogg, A. 1991, Personal communication. Wild Hill Orchard. Ely, VT

Fowler, C. and Mooney, P. 1990, Shattering: Food, Politics, and the loss of Genetic Diversity. The University of Arizona Press. Tucson, AR

Grainge, M. and Ahmed, S. 1988, Handbook of Plants with Pest Control Properties. John Wiley & Sons. New York, NY

Heinrichs, E.A. (Editor) 1988, Plant Stress-Insect Interactions. John Wiley & Sons, Inc. New York, NY

Hillel, D.J., 1991, Out of the Earth: Civilization and the Life of the Soil. The Free Press. New York, NY

Leius, K. 1967, Influence of wild flowers of parasitism of tent caterpillar and codling moth. Can. Entomol. 99: 444-446

Parnes, R. Fertile Soil: A growers Guide to Organic & Inorganic Fertilizers. Ag Access, Davis, CA

Price, P.W., Lewinsohn, T.M., Fernandes, G.W. and Benson, W.W. (Editors) 1991, Plant - Animal Interactions. John Wiley & Sons, Inc. New York, NY

Senn, T.L. 1987, Seaweed and Plant Growth. No publisher noted. Department of Horticulture, Clemson University. Clemson, SC

Seyer, E. 1992, Sustaining a Vermont Way of Life: Research and education in Sustainable Agriculture. University of Vermont. Burlington, VT

Smith, C.M. 1989, Plant Resistance to Insects, A fundamental Approach. John Wiley & Son, Inc. New York, NY

Talbot, M. 1992 Personal Communication. Dorchester, MA

Walters, C. Jr. 1991, Weeds: Control Without Poisons. Acres USA. Kansas City, MO

Waksman, S.A. 1936, Humus. Williams and Wilkins, Inc. Baltimore, MD

Ziv, O. and Zitter, T.A. 1991, Effects of Bicarbonate and Coating Materials on Cucurbit Foliar Diseases. Proceedings of the 53rd Annual Pest Management Conference, Nov. 11-14, 1991. Dept. of Plant Pathology, Cornell Univ. Ithica, NY

Part II
IN PRACTICE

Chapter 5
COMPOSTING
and
PRESERVING
ORGANIC MATTER

Organic matter (OM) is an essential component of a natural soil system. It would be hard to find any expert who disagrees with this statement. Some of the benefits of soil organic matter (SOM) include:

Increased water holding capacity - organic matter is like a sponge. It can hold four to six times its own weight in water (see chapter 2). The ability of the soil to hold more solution also increases its capacity to retain more dissolved nutrients.

Improved soil structure - OM binds soil particles together for 1) better retention of water and nutrients, and 2) it decreases the potential of erosion. It can also aggregate clay soil by interrupting the bonds between clay particles that cause compaction and porosity problems (see chapter 2).

Increased nutrient holding capacity - humus is a colloidal substance that holds onto mineral nutrients such as potassium, magnesium and calcium (see chapters 2 & 7). Nutrients such as nitrogen, phosphorus and sulfur exist as proteins and other carbon compounds in OM and can be mineralized into available plant food by bacteria at a slow and sustained rate.

Increased microbial populations - microbes thrive in organic matter. It has everything the little critters need to exist including water, food and air. Consequently, many different varieties of micro-organisms are constantly competing for those available resources. That competition creates a biological equilibrium that, in the long and short term, can prevent many insect, disease, and even nutrient imbalance problems that plague the grower and land care expert (see chapter 4). The carbon in organic matter is *the* source of energy for most micro-organisms. The greater the available source of energy, the larger and more diverse their populations can be.

Reduced soil density - organic matter is lighter than sand, silt or clay; a real benefit to root growth. As roots travel through the soil, energy, manufactured by photosynthesis, is utilized. The more difficult the journey, the more energy the plant uses to travel the same distance and the less root growth is ultimately obtained. As roots travel and branch out in the soil, they gain access to more nutrients needed to synthesize more energy.

Improved mineral nutrient availability - some mineral nutrients are made available in the soil from the activities of organisms on parent material (rock particles). There is approximately 52,000 pounds of potash, 40,000 pounds of magnesium, 72,000 pounds of calcium, 3,000 pounds of phosphate, 3,000 pounds of sulfur and

MINERAL COMPONENTS
8 Elements Comprise 98% of All Soil Mineral

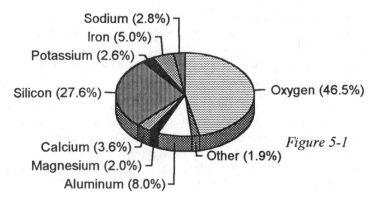

Sodium (2.8%)
Iron (5.0%)
Potassium (2.6%)
Silicon (27.6%)
Oxygen (46.5%)
Calcium (3.6%)
Magnesium (2.0%)
Aluminum (8.0%)
Other (1.9%)

Figure 5-1

100,000 pounds of iron in an average acre of topsoil (see figure 5-1). Albeit abundant, all of these nutrients are bound in mineral structures that are insoluble and unavailable to plants. Organisms in nature can dissolve some of these minerals into available nutrients by creating enzymes and acids that are corrosive to the surface of the rock particles. Many organisms, including plants, produce these corrosive exudates specifically to liberate mineral nutrients for their own metabolism. Their existence and proliferation is directly influenced by soil organic matter (see chapter 1).

Increased nitrogen efficiency - nitrogen is used by plants and other organisms to produce protein. When they die and are returned to the soil as OM, that complicated process which created the protein is essentially reversed by soil bacteria, and it is converted back into mineral nitrogen. The bacteria utilize some of the nitrogen for their own protein synthesis and make some available to plants. Eventually, the bacteria die and their protein is converted into available nitrogen by other bacteria and more is made available to plants. The beauty of this system is the incredible efficiency of nitrogen utilization (see chapter 2).

Increased porosity - humus itself is not porous, but its ability to hold as much water as it does can create porosity. Humus swells in order to accomodate as much as four to six times its own weight in water. This swelling causes a sort of heaving in the soil. When water is released during dryer periods, the OM contracts and leaves behind air spaces. Earthworms and other larger organisms that burrow through the soil looking for organic matter to digest create even more porosity.

The list of benefits from OM can go on for several hundred pages, and the expense of providing those benefits without OM is staggering. In our routine cultural practices, many of us are inadvertently depleting soil organic matter content. When organic matter levels begin to drop, so do all the corresponding benefits, including water and nutrient holding capacity and microbe populations and their activities. This loss has a profound effect on the soil, creating a type of economic recession for organisms (including plants). The only things increased by lower levels of OM are bulk soil density and problems.

BUILDING OM

Building organic matter is not as easy as it may seem. The decomposition of OM is an integral part of the carbon cycle (see Figure 5-2). Plants would be unable to produce protein, carbohydrates and other components of their tissue if carbon dioxide were not evolved from OM and recycled back into the atmosphere.

CARBON CYCLE

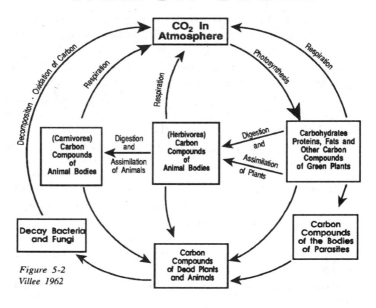

Figure 5-2
Villee 1962

Adding OM to increase soil levels significantly is a monumental task. If an acre of topsoil (6-7 inches deep) weighs two million pounds, then increasing OM content by one percent would require 20,000 pounds (10 tons) of stable humus (approximately 40 - 50 cubic yards).

Green manure crops, such as clover or vetch, can contribute significant amounts of OM to the soil, but it may be impractical for certain applications. It is unlikely that a home owner or golf course superintendent, for example, would turn his turf into a green manure crop for a year or two just to build soil organic matter.

Adding compost will also stimulate many beneficial soil functions, but compost and green manures are not necessarily stable humus. Only 1-10% of these inputs may actually become stable soil OM (depending on soil and climate conditions) and in some cases, less. In many situations where organic matter is very depleted, adding compost may be an absolute necessity. However, repeated applications on an annual basis may be required to slowly replenish the soil's OM reserve.

If cultural practices initially caused the depletion of SOM and are not changed, then the problem will eventually occur again. It is necessary to be aware of the conditions that accelerate the depletion of organic matter in the soil, to try to slow down the process, and to preserve what is already there.

COMPARISONS

To illustrate the factors controlling the existence of OM, a comparison is made of: 1) the art of composting and 2) the maintenance of SOM. Although the objectives of both activities are similar (i.e. creating and maintaining more OM in the soil) the methods of accomplishing each are opposite. Understanding how to make compost is important because the same forces that decay organic matter in a compost pile will also deplete it in the soil. Knowing the best way to make compost quickly can provide insight on how to slow down or accelerate that same process in the soil. In agriculture, it is often necessary to speed up the decomposition of SOM to release the organically bound nitrogen.

COMPOSTING

Composting is as much an art as it is a science. Learning it takes some effort and time but, like anything else, practice provides proficiency. Keep in mind that the decay of organic matter is a natural phenomenon and will occur with or without our help. All we are trying to do is speed up the process.

There are five factors needed to accelerated the decay process to successfully compost organic materials in a reasonable amount of time. These same factors must be limited in cultivation to preserve soil organic matter (SOM). The factors are:

Air
Water
Carbon:Nitrogen (C:N) Ratio
Temperature
pH

AIR

In the compost pile:

The bacteria that are most adept at breaking down organic matter are aerobic organisms. They, like us, need oxygen to live. If a pile of material is wet and compressed, there will not be enough air for the right type of bacteria and another type, the anaerobic bacteria, will process the pile. This situation will cause odor problems and will not create the heat necessary to kill weed seeds and pathogens. Turning and aerating the material several times during the composting process will increase the amount of oxygen in the pile. Many compost operations are set up on top of perforated pipes to insure that plenty of air is reaching the bottom of the pile. There is new research being conducted based on the theory that proper mixing and a piped air infiltration system are all that is needed for proper aeration. The heat generated by the pile creates a chimney effect that draws fresh air in through the pipes. No turning or motor driven blowers are necessary. However, in conventional composting methods, turning is a crucial component.

Knowing when and how often to turn a pile is the key to making a finished compost quickly. Unfortunately, there is no set formula. The number and frequency of turns necessary changes with the climate and the different materials being decomposed. A good method of determining turning times is using a thermometer with a two to four foot long probe to monitor the temperature of the pile. When the pile is first mixed the temperature should rise rapidly to between 120 and 140 degrees. Within a week or two the temperature should drop. When the decrease exceeds twenty degrees below its peak, it is time to turn the pile. Heat will increase again rapidly each time the pile is turned until the composting process is completed, at this point the compost should be ready to use. Eventually, familiarity with the composting process may eliminate the need for a thermometer.

In some large composting operations, wind-row turners are utilized to mix and aerate the pile. These machines are specifically designed to add air into the wind-rows of OM being composted.

Surface area is another aspect of aeration that should be considered. The smaller the particle of organic material, the more surface area is exposed to decay bacteria, and the faster it is decomposed. Saw dust decays faster than chips, which decompose considerably faster than whole branches. A proper mixture of coarse and fine materials also helps air infiltration. Material that is too fine can decrease air infiltration through the pile and cause anaerobic conditions.

In the soil:

The bacteria that decompose soil organic matter (SOM) most actively are also aerobic. However, there arc thousands of other beneficial soil organisms that are aerobic; therefore depriving the soil of oxygen is not being suggested here. It is important to evaluate current cultural practices and to determine if the amount of oxygen being introduced into the soil is excessive.

Plows, rototillers, harrows and other cultivation equipment accomplish essentially the same type of aeration as the turning tools used for composting. Their task is to lighten the soil by incorporating more air into it. The supposed need for this activity is to create a favorable environment for seeds or seedlings where roots can easily move through the soil; to provide them access to oxygen (a necessary nutrient); and to incorporate surface residues into the soil, to accelerate the time it takes for their decay and the recycling of the nutrients they contain.

Unfortunately, too much of this activity can set in motion a chain reaction that depletes the level of SOM and a related decrease in the benefits it provides. As SOM slowly disappears, the soil becomes heavier and requires more cultivation, which increases the rate of decay even further. The dust bowl days of the 1930's American midwest are an extreme example of what over tillage can cause.

Research in no-till or low-till programs is discovering that a significant amount of SOM can be preserved just by reducing the amount of air introduced into the soil.

Core aeration is a conventional practice in turf maintenance that ostensibly introduces air and reduces soil density, providing a better environment for turf roots. However, like the plow, the over-use of this tool can reduce SOM and increase the need for its continued and more frequent use. An application of compost after core aeration can mitigate the effects of oxidizing SOM.

It is important to evaluate the need for aeration and reduce the indiscriminate use of these tools. Sandy soils are naturally aerated because of the course particle size. Tillage or core aeration in these types of soils can often be reduced or even eliminated. High traffic areas that are prone to compaction need aeration from time to time. However, a soil building program should be implemented to counter the effects of aeration. Turf, for example, could be topdressed with compost after core aeration at least once or twice a year.

Soil compaction is a reason for aeration but is often caused by a reduction of SOM from aeration. A simple test for compaction is to pour a container of water onto a given area and observe how quickly it seeps into the soil. This test should be performed on soils that are already slightly moist (super dry surfaces can actually repel water in some cases). Fast absorption usually means plenty of porosity. In extremely sandy conditions, some managers have successfully decreased soil porosity by adding colloidal phosphate, which contains a natural clay that binds some of the sand particles together. However, it should only be used where soil phosphate levels are moderate to low. Other commercially available clays have been used with mixed results. If excessive use of aeration equipment is ultimately the cause of compaction, then further use is a treatment of the symptom and inevitably makes the problem worse.

WATER

In the compost pile:

All living things need water to live, and the organisms that make compost are no different. A compost pile should contain somewhere between forty and sixty percent moisture. A simple test to determine proper moisture content is to squeeze a handful of raw material. The material should feel damp to the touch but no more than a couple of drops of water should be expelled. If the pile is too dry, bacterial reproduction will be inhibited, which will slow down the composting process. If the pile is too wet it can become anaerobic. Adding water to a pile is easy enough if it is needed, but extracting it is impractical. To correct excess moisture content dry, bulky materials need to be added to the pile to absorb the water. Saw chips or dust, planer shavings or shredded newspaper can be used. Dry leaves are suitable if it is the season for them, and soil can also be used in a pinch. Turning (aerating) the pile can also help dry it out.

In the soil:

Water can, in many ways, be both friend and enemy. Too much or too little water can cause serious problems for anyone involved in growing or maintaining plants. Bacteria that decompose organic matter are as dependent on moisture as plants are or any other living thing. As in the compost pile, these organisms do not do well during periods of extreme dryness or saturation. If water is supplied through irrigation, it is important to monitor soil moisture levels carefully and to practice moderate watering techniques to encourage root growth (large producers of OM). New research suggests that low volume, high frequency watering techniques can improve plant and soil health while using water more efficiently. Other experts disagree and suggest that deep and infrequent watering is best for most plants. Common sense dictates that moisture conditions that are ideal for plants are most often ideal for the accumulation of OM, simply because OM is being produced faster by plant growth than it is being decomposed by bacteria. What the ideal level is varies in different soils and is discussed in Chapter 3. Indiscriminate applications of water can contribute to the accelerated demise of SOM. In most cases, the supply of water is controlled completely by climate and cannot be adjusted. The balance between air and water in the soil provides a natural regulation of SOM decay. Periods of excess water deplete the amount of air, inhibiting the activities of decay organisms. During dry periods, air is abundant, but moisture becomes the limiting factor. Maintaining adequate levels of SOM can buffer the effects of too much or too little rain.

TEMPERATURE

In the compost pile:

If all the other conditions necessary for proper composting exist, temperature will come naturally. Pile temperatures between 90 and

140F degrees are necessary for decomposition to occur in the shortest amount of time. Below ninety degrees composting slows significantly, while temperatures above 140 are too high for most microbes to live. If the pile is too cold, chances are one of the other conditions such as air, moisture or C:N ratio are at improper levels. It can also mean that the pile is not big enough to insulate itself from colder outside air. A minor adjustment of one of these components can heat up a pile quickly. Piles that are too hot can be self regulating. If the microbes cannot live in temperatures above 140 degrees then the pile should automatically cool down when they die. However, on some occasions a pile may have to be manually cooled down with water or aeration to regain normal temperatures. Proper pile size and configuration can control overheating. Piles can, and do, heat up during the winter months, but only if they have sufficient mass to insulate themselves from the cold air temperature.

To conserve heat, waste materials need to be concentrated. However, piling materials too high can cause overheating and odors. If the height of your pile exceeds ten feet, then conversion into a windrow is advised. The windrow pile can accommodate a tremendous volume of material providing that the size of the site is adequate.

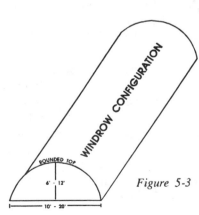

Figure 5-3

The windrow shape should be rounded with a base approximately ten to twenty feet wide with a height of about six to twelve feet. The top can be made concave to catch rain water if the pile is too dry, otherwise it should be rounded to shed water. The overall length is determined by the amount of material being processed (see figure 5-3). Small composting vessels designed for home use work well in the warmer seasons, but cannot generate enough heat to function during freezing winter temperatures.

In the soil :

Temperature can have a profound effect on the rate of SOM

Influence of
Soil Temperature
on Humus Accumulation

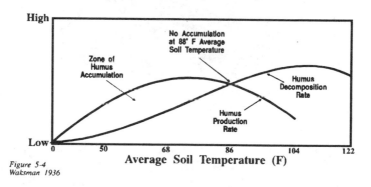

Figure 5-4
Waksman 1936

Average Soil Temperature (F)

decomposition. As illustrated in Figure 5-4, at a soil temperature of 88° F (31.1°C), with adequate air and moisture, SOM is destroyed faster than it can be produced. This is a common condition in tropical soils where high temperature, moisture from tropical rains, and the abundance of air from extremely sandy soil are all at an optimum level for decay bacteria. There is a direct correlation between the average annual temperature of a given region and the native levels of SOM. As one moves closer to the equator, it is evident that the natural existence of SOM lessens (see figure 5-5).

Influence of Temperature
on Soil Organic Matter Content

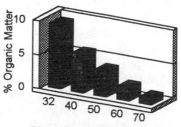

Figure 5-5
Jenny 1941

Mean Annual Temperature

Influence of Mulch
on Soil Temperature

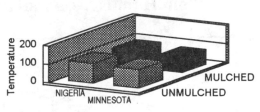

Figure 5-6
SSSA# 19

Location

The effects of high temperatures on SOM can be mitigated by shading the soil. Straw and other organic mulches can lower the temperature of the soil by as much as 20°F (9.4°C). Living mulches can also provide protection from the soil's absorption of the sun's energy. Leaving bare soil exposed to the sun can cause a significant increase in soil temperature (see figure 5-6). For example, turf stands that are mowed at maximum height during the hottest part of the year can shade the soil and reduce soil surface temperature. This practice is not always possible because of use restrictions such as on golf greens and ball fields. However, in many instances a tall green stand of turf is much preferred over a short brown one. Allowing turf to stand a little taller not only provides a natural shade for the soil but also encourages a greater productions of OM (root systems can contribute from 0.5-4 tons of OM per acre per year).

NITROGEN

In the compost pile:

The ratio of carbon to nitrogen (C:N) is important when combining materials for composting. The ideal C:N ratio for a compost pile is between twenty five and thirty five parts carbon to one part nitrogen (average 30:1). If the C:N ratio of your pile is too high (i.e. high carbon, low nitrogen) composting will be slow because the amount of available nitrogen will be too low for proper protein synthesis, and the reproduction of decay organisms will be inhibited. If the C:N ratio is too low, (i.e. too much nitrogen) reproduction can be over-stimulated causing oxygen depletion and, eventually, anaerobic conditions. If there is insufficient carbon for microbial processing much of the excess nitrogen will be lost to leaching or volatilization. Estimating how much of what material to add to the pile can be accomplished by using some of the values below (see table 5-1):

Most of the values given in table 5-1 are estimates because the C:N ratio of organic residues can vary depending on age, variety, where it is raised, or what it is raised on. C:N values are calculated on a dry matter basis, but compost mixes are usually measured by volume. So developing a mathematical formula to determine exactly how much of each component to use is not practical. However, it is not all that critical. Looking at table 5-1 gives a rough idea of what to use to either raise or lower the C:N ratio of a pile.

table 5-1

WASTE	C:N RATIO
Grass clippings	9-25:1
Leaves (green)	30-50:1
Leaves (autumn)	50-80:1
Wood and sawdust	300-700:1
Paper	150-200:1
Bark	116-1285:1
Straw	48-150:1
Cow manure	11-30:1
Horse manure	22-50:1
Hen manure	3-10:1
Sheep manure	13-20:1
Food wastes	14-16:1

The ideal level for fast decomposition is around 30:1 so a combination of materials that have values above and below that level would be a good starting place. Depending on actual ingredients, particle size and moisture content, small adjustments may be necessary. Like anything else, experience makes it easier.

If the majority of available wastes have a high C:N ratio, organic or inorganic nitrogen fertilizer can be used to create a better balance. However, it should be used very judiciously and mixed in thoroughly. Remember, if a little is good, more is not necessarily better. Reducing particle size of the waste by grinding or shredding can help speed up the process too.

In the soil:

Nitrogen is a nutrient that is vital to the development of decay bacteria populations as well as the growth of plants. The ratio of carbon to nitrogen in an average soil is approximately 12:1. At this ratio, populations of decay bacteria are kept at a relatively constant level. However, when large quantities of inorganic N are added to the soil, the ratio is changed and the bacteria populations are stimulated to do what they do best...decompose organic matter. Stable humus is relatively resistant to further decomposition even when bacteria are stimulated by added nitrogen. Younger, less evolved material however, is not and may never evolve into stable humus if subjected to conditions that are conducive to quick and complete decomposition (see chapter 2). An increase in bacterial activity can also cause a depletion of soil oxygen levels which can inhibit root growth, slowing production of OM and increasing stress to the plant.

Natural organic nitrogen can mitigate this problem because it is organic matter and provides some of the energy (in the form of carbon compounds) needed by decay organisms. However, excess applications of nitrogen in any form can begin to cause problems.

Judicious use of inorganic nitrogen applied along with organic amendments can be more effeciently used by the high populations of bacteria that exist in soils with adequate levels of OM. These organisms can utilize inorganic nitrogen for their own protein production thus, making more efficient use of it. Once in a protein form, nitrogen is not vulnerable to leaching or volatilization so losses are minimized. Unfortunately, too much inorganic nitrogen at any one time applied without any organic carbon can cause bacteria to decompose existing organic matter at an alarming rate, possibly causing the destruction of their own habitat (see chapter 2).

As SOM is depleted, so is the efficiency of inorganic nitrogen utilization which creates the need for more frequent and larger applications. This, of course, makes the problem even worse.

Effects of pH on Soil Life

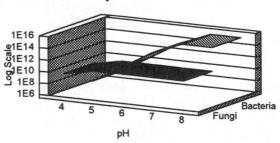

Figure 5-7
Waksman 1936

pH

In the compost pile:

The pH of the material to be composted is important to consider. Materials such as pine needles or oak leaves that have a low pH can inhibit the growth of the decay bacteria (see figure 5-7). Alkaline materials like manures can cause chemical reactions that give off strong odors. However, mixing high and low pH items together, will result in a near neutral pH, while stabilizing much of volatile materials that can cause odors. If the pH of a pile is too low, lime or ash can be mixed in to raise it. Figure 5-7 shows the biological reaction of bacteria to changes in pH.

In the soil:

Indiscriminate applications of lime can raise the soil pH to levels that are too favorable for decay organisms (see figure 5-7). The resulting increase in their population can cause a related decrease in SOM. Professionals and home owners who lime without the benefit of a soil test run the risk of creating this scenario. In figure 5-8, it is evident that increased levels of lime also increase the evolution of carbon dioxide from the decomposition of soil organic matter. A soil pH test should be the only criterion for applying lime.

Influence of Lime
on Organic Matter Decomposition

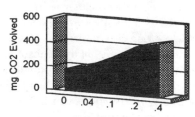

Figure 5-8
Waksman 1936

Percent Calcium Carbonate

MOUNT MANSFIELD

On the summit of Mt. Mansfield (Vermont's highest peak) the soil is very rich in organic matter. Why such an inhospitable place should have high SOM is interesting to note but the reasons are simple:

1. **Air** - the soil is undisturbed, not cultivated for any purpose; therefore, soil aeration is kept at a minimum.

2. **C:N ratio** - the residues are almost exclusively from trees, contributing only small amounts of nitrogen to the decay process.

3. **Temperature** - average annual temperatures on the summit are well below the those of the surrounding region. Snow covers the peak for as many as nine months per year.

4. **pH** - both high levels of precipitation and organic residues inherently low in base cations contribute to a very low pH on the mountain.

All of these factors inhibit the activities of decay organisms and allow for accumulations of SOM well above native levels in the surrounding lowlands. Creating an environment inhospitable to decay organisms is not what is being suggested here. The objective is to understand the balance between the plant, the soil and the organisms that inhabit it.

HOW MUCH SOM IS ENOUGH

OM content can be measured by most soil testing labs using the same samples submitted for nutrient analysis (see chapter 7). The results of the test are usually expressed as a percent of soil content. Five percent or more is an excellent level to have but is not always practical to obtain. Soils located closer to the equator tend to have lower levels of SOM as a result of average soil temperatures (see figure 5-5). Under certain conditions, building OM levels to five percent might be impossible (e.g. in tropical soils). However, any attempts toward improving OM levels will usually cause an improvement in overall soil conditions. Soils in temperate regions of the world that contain less than two percent SOM may be nearing a critical level and need evaluation.

Questions such as the following, may provide clues to determining if a SOM building program is necessary: Do the crops in this soil have excessive problems with compaction or water retention? Do signs of drought arrive in this location earlier than anywhere else? Or, is the incidence of fungal disease or insect attack more frequent in this spot? If the answer to any of these questions is *YES*, then the level of OM may have been depleted below the threshold necessary for a functioning, natural soil system.

SUMMARY

It is important to realize that organic matter is an asset and needs to be managed like anything else of value. And, like most other assets, organic matter is much easier to maintain than it is to replace. The development of organic matter from compost is appropriate technology for sustaining SOM levels and proper waste manage ment. Land applications of composted wastes can improve fertility, but accumulation of SOM is not necessarily accomplished without changing the cultural practices that originally depleted it. The natural resources stored in organic wastes must be recycled or eventually all the wealth of the planet will be found in the landfills.

INFORMATION ON COMPOSTING

It is important to understand that proficiency in composting is like learning to tie one's shoes; confusion and lack of confidence will exist the first few times it is done. Think small at first. If one can develop a technique to heat up a bushel of waste, it can used on a thousand cubic yards.

This chapter provides only an overview of composting. It may not contain some of the details needed to compost organic wastes on a larger scale. To investigate the subject in more detail, a list of appropriate books is provided:

YARD WASTE MANAGEMENT: A PLANNING GUIDE FOR NEW YORK STATE. Prepared by T. Richard, N. Dickson and S. Rowland. Dept. of Agriculture and Biological Engineering, Cornell University. Ithaca, NY 14853-5701. 163 pp.

THE BIOCYCLE GUIDE TO YARD WASTE COMPOSTING 1989. The JG Press, Inc. Emmaus, PA 18098. 197pp

COMPOSTING: A STUDY OF THE PROCESS AND ITS PRINCIPLES. C. G. Golueke, 1972. Rodale Books, Inc. Emmaus PA 18049

THE BIOCYCLE GUIDE TO THE ART AND SCIENCE OF

COMPOSTING. 1991 The JG Press, Inc. Emmaus, PA 18098. 270pp

LET IT ROT: THE GARDENERS GUIDE TO COMPOSTING. S. Campbell. 1975. Garden Way Publishing. Storey communications, Inc. Pownal, VT 05261. 152 pp.

THE RODALE BOOK OF COMPOSTING. D. Martin and G. Gershuny 1992 Rodale Press, Emmaus, PA 18049.

HOME COMPOSTING: A TRAINING GUIDE. N. Dickson, T. Richard, B. Kozlowski and R. Kline. 1990. NRAES, Riley Robb Hall, Cornell University, Ithaca, NY 14853

ON-FARM COMPOSTING HANDBOOK. Northeast Regional Agricultural Engineering Service 1992. NRAES, Riley Robb Hall, Cornell University, Ithaca, NY 14853. 186pp

Sources:

Albrecht, W.A. 1938, Loss of Organic matter and its restoration. U.S. Dept. of Agriculture Yearbook 1938, pp347-376

BioCycle Staff (editors) 1989, The Biocycle Guide to Yard Waste Composting. JG Press, Inc. Emmaus, PA

BioCycle Staff (editors) 1990, The Biocycle Guide to Collecting, Processing and Marketing Recyclables: Including the Special Report on Materials Recovery Facilities. JG Press, Inc. Emmaus, PA

BioCycle Staff (editors) 1991, The Art and Science of Composting. JG Press, Inc. Emmaus, PA

Buchanan, M. and S.R. Gliessman 1991, How Compost Fertilization Affects Soil Nitrogen and Crop Yield. Biocycle, Dec. 1991. J.G. Press Emmaus, PA

Golueke, C.G. 1972 Composting: A Study of the Process and its Principles. Rodale Books, Inc. Emmaus, PA

Holland, E.A. and Coleman, D.C. 1987. Litter Placement Effects on Microbial and Organic Matter Dynamics in an Agroecosystem. Ecology v68 (2), 1987: 425-433

Lucas, R.E. and Vitosh, M.L. 1978, Soil Organic Matter Dynamics. Michigan State Univ. Research Report 32.91, Nov 1978. East Lansing, MI

Makarov, I.B. 1986, Seasonal Dynamics of Soil Humus Content. Moscow University Soil Science Bulletin, v41 #3: 19-26

Nosko, B.S. 1987, Change in the Humus for a Typical Chernozem caused by fertilization. Soviet Soil Science, 1987 v19 July/August p67-74

Novak, B. 1984, The Role of Soil Organisms in Humus Synthesis and Decomposition. Soil Biology and Conservation of the Biosphere. pp 319-332

NRAES, 1992, On Farm Composting Handbook. Northeast Regional Engineering Service #54, Cooperative Extension. Ithica, NY

Parnes, R. Fertile Soil: A growers Guide to Organic & Inorganic

Fertilizers. Ag Access, Davis, CA

Parr, J.F., Papendick, R.I., Hornick, S.B. and Meyer, R.E. 1992, Soil Quality: Attributes and relationship to alternative and sustainable agriculture. American Journal of Alternative Agriculture v7 #1 and 2, 1992 pp 5-10. Institute for Alternative Agriculture, Greenbelt, MD

Richards, T., Dickson, N. and Rowland, S. 1992 Yard Waste Management: A planning Guide for New York State. Dept. of Agriculture and Geological Engineering, Cornell University. Ithaca, NY

Seyer, E. 1992, Sustaining a Vermont Way of Life: Research and education in Sustainable Agriculture. University of Vermont. Burlington, VT

Silkina, N.P. 1987, Effects of High Nitrogen Fertilizer Concentrations on Transformation of Soil Organic Matter. University of Moscow Soil Science Bulletin 1987, v42 (4): 41-46.

Singh, C.P. 1987, Preparation of High Grade Compost by an Enrichment Technique. I. Effect of Enrichment on Organic Mater Decomposition. Biological Agriculture and Horticulture 1987, vol 5 pp 41-49

Smith, G.E. 1942, Sanborn Field: Fifty Years of Field Experiments with Crop Rotations, Manures and Fertilizers. University of Missouri Bulletin #458. Columbia, MO

SSSA# 19. 1987, Soil Fertility and Organic Matter as Critical Components of Production Systems. Soil Science Society of America, Inc. Madison, WI

Veen, A. van and Kuikman, P.J. 1990. Soil structural aspects of decomposition of organic matter by micro-organisms. Biogeochemistry, Dec. 1990, v11 (3): 213-233

Villee, C.A. 1962. Biology. W. B. Saunders Company. Philadelphia, PA

Visser, S. and Parkinson, D. 1992, soil biological criteria as indicators of soil quality: Soil microorganisms. American Journal of Alternative Agriculture v7 #1 and 2, 1992 pp 33-37. Institute for

Alternative Agriculture, Greenbelt, MD

Wallace, A., Wallace, G.A. and Jong, W.C. 1990. Soil Organic Matter and the Global Carbon Cycle. Journal of Plant Nutrition 1990 v13 (3/4): 459-456

Waksman, S.A. 1936, Humus. Williams and Wilkins, Inc. Baltimore, MD

Waksman, S.A. and Woodruff, H.B., The occurrence of bacteriostatic and bactericidal substances in the soil. Soil Science v53 pp223-239.

White, W.C. and Collins, D.N. (Editors) 1982, The Fertilizer Handbook. The Fertilizer Institute. Washington, DC

Chapter 6

ORGANIC VS INORGANIC

Organic does not mean natural, nor does it necessarily mean good. Technically, the definition of organic refers to a complex of chemical bonds between three elements: carbon (C), hydrogen (H), and oxygen (O). Traditionally, the word organic has meant anything that contains carbon compounds derived from living organisms. These different interpretations cast a shroud of confusion upon the world of agriculture and horticulture. Synthetic pesticides, for example, are almost all organic by the technical definition, but are prohibited by organic practitioners that certify their products as *organically grown*. Fertilizers that contain components such as bone meal or rock phosphate cannot be considered all organic because those ingredients do not contain organic carbon. However, both items are allowed for certified organic food production. Urea, which is a synthetic form of nitrogen, is technically organic because of its carbon, hydrogen, and oxygen content, but it is prohibited by organic growers.

Defining organic when referring to fertilizers might be better accomplished by adding the word natural. However, that term is also misconstrued by some consumers and misrepresented by some manufacturers. A natural organic fertilizer should be defined as a product that contains ingredients derived from plants or animals that have not been chemically changed by the manufacturer. Materials that fit into this definition are derived from plant and animal residues.

Many manufacturers add natural minerals such as rock phosphate or potash salts to enhance the nutritive value of their products. These ingredients are acceptable for use by organic growers, but they are not organic. Natural, inorganic amendments not only change the fertilizer but also, in most parts of the United States, change the way in which the product must be labeled. A manufacturer can no longer label the fertilizer products as *Natural Organic* because the mineral ingredients are not organic.

Figure 6-1 shows the relationship between the four terms most commonly used to define inputs. Both natural-organic and synthetic-inorganic are relatively commonplace terms and well understood. But there are many examples of natural inorganic and synthetic organic materials. Lime, natural potash, phosphate rock and Chilean nitrate are all examples of natural inorganic materials. Urea is a well known example of synthetic organic.

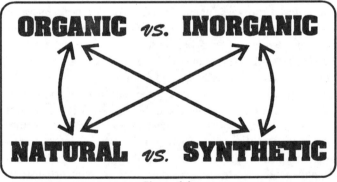

Figure 6-1

Categorizing any of these materials as good or bad based solely on the terms in figure 6-1 is inappropriate. Judgment should be based on the physical and biological impact each material has on the soil. Unfortunately, many groups involved with certifying produce or land care practices as *organic* add a marketing factor to the equation, obscuring the original intent of the certification process. The marketing component results from public perception and, wrong or right, it is not easily changed.

This chapter attempts to analyze each elemental plant (and soil) nutrient, and to review some of the materials it is are derived from. Emphasis is applied to those materials that are allowed by most committees involved in organic certification. Table 6-1 shows a

Table 6-1

PRODUCT	ORGANIC	INORGANIC	NATURAL	SYNTHETIC	ALLOWED
Blood meal	Yes	No	Yes	No	Yes
Feather meal	Yes	No	Yes	No	Yes
Leather meal	Yes	No	Yes	No	Some
Vegetable Protein meal	Yes	No	Yes	No	Yes
Animal tankage	Yes	No	Yes	No	Yes
Dried whey	Yes	No	Yes	No	Yes
Fish meal	Yes	No	Yes	No	Yes
Natural nitrate of soda	No	Yes	Yes	No	Restricted
Urea	Yes	No	No	Yes	No
Rock phosphate	No	Yes	Yes	No	Yes
Black rock phosphate	No	Yes	Yes	No	Yes
Colloidal rock phosphate	No	Yes	Yes	No	Yes
Calcined rock phosphate	No	Yes	Yes	No	Yes
Soap phosphate	No	Yes	Yes	No	Yes
Raw bone meal	Yes	Yes	Yes	No	Yes
Steamed bone meal	Yes	Yes	Yes	No	Yes
Precipitated bone meal	No	Yes	?	?	Yes
Precipitated milk phosphate	No	Yes	?	?	Yes
Super phosphate	No	Yes	No	Yes	No
Triple super phosphate	No	Yes	No	Yes	No
Ash	No	Yes	Yes	No	Yes
Greensand	No	Yes	Yes	No	Yes
Sulfate of potash	No	Yes	Yes	No	Yes
Sulfate of potash, magnesia	No	Yes	Yes	No	Yes
Muriate of potash	No	Yes	Yes	No	No
Trace elements	No	Yes	No	Yes	Restricted
Manure	Yes	No	Yes	No	Restricted
Compost	Yes	No	Yes	No	Yes
Green manures	Yes	No	Yes	No	Yes
Kelp meal	Yes	No	Yes	No	Yes
Seaweed Extract	Yes	No	Yes	No	Yes
Humates	Yes	No	Yes	No	Yes
Beneficial bacteria	Yes	No	Yes	No	Yes
Ground limestone	No	Yes	Yes	No	Yes
Aragonite	No	Yes	Yes	No	Yes
Gypsum	No	Yes	Yes	No	Yes
Epsom salts	No	Yes	Yes	No	Restricted

comparison of different fertility elements and how they relate to the four terms outlined in figure 6-1. Because most certification groups differ slightly in what they **allow**, the column so marked is based on an average or majority.

NITROGEN

For years land managers, from the farmer to the scientist, have disputed the benefits of organic versus inorganic nitrogen. There is no dispute that nitrogen is an essential element to plants. There is also no argument that plants can't tell the difference between organic and chemical nitrogen. The controversy is essentially about carbon.

Carbon and nitrogen react to each other a little like siblings. In plants, they function together to form amino acids, enzymes and proteins. In the soil, they antagonize each other if they get out of balance. Too much carbon can immobilize all of the available nitrogen, and excess nitrogen can deplete soil carbon.

Carbon in the soil is in the form of organic matter and provides energy, either directly or indirectly, to all heterotrophs (i.e. living organisms that utilize carbon compounds directly from plants and other organisms). Soil carbon is produced by autotrophic organisms such as plants and algae that can fix carbon from the atmosphere by utilizing energy from the sun. The carbon compounds produced by autotrophs eventually become part of a vast warehouse of energy and protein known as soil organic matter. This warehouse functions beneficially in hundreds of different ways, but one essential purpose is to provide energy to soil life.

When fresh organic matter (OM) hits the soil, decay begins almost immediately (i.e. during the seasons that micro-organisms are active). What determines the speed at which OM is decomposed (with adequate air and moisture) is the carbon:nitrogen ratio (C:N) of the OM. The C:N ratio is always measured as x parts carbon to one part nitrogen. If the C:N ratio is high (i.e. high carbon) such as in straw or wood chips, decomposition occurs slowly. Also, the nitrogen is temporarily commandeered by bacteria from other sources for the formulation of proteins. If the original organic litter has a low C:N ratio, such as grass clippings or animal wastes, decomposition will occur more rapidly and nitrogen is made available to other organisms. Each time the components of OM are digested by heterotrophs some energy is utilized, and carbon is oxidized into carbon dioxide (CO_2) that is released back into the atmosphere.

CARBON CYCLE

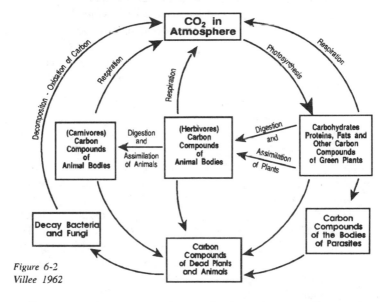

Figure 6-2
Villee 1962

Figure 6-2 shows how carbon cycles from the atmosphere to the soil and back into the atmosphere. Carbon dioxide in the atmosphere is absorbed by plants and transformed into carbohydrates, proteins and other organic compounds. These compounds are essentially storage batteries containing energy that was originally derived from the sun. If the plant is consumed by animals, some of the original energy is used and CO_2 is released back into the atmosphere. If the animal is consumed by another animal, more of the energy is utilized and more CO_2 is released. Eventually the remaining energy is returned to the soil in the form of animal residues where decay organisms can utilize it. If plant residues are introduced directly to the soil without prior consumption, more energy (i.e. carbon) will be available to soil microbes.

Nitrogen (N) serves the microbe as much as (or more than) it serves the plant. If there is only enough nitrogen in the soil for either the plant or the needs of bacteria, the bacteria will get it.

When inorganic nitrogen is applied to the soil it stimulates populations of decay bacteria as well as plants. If used judiciously, it can have a synergistic effect with OM that increases overall nitrogen efficiency. Large populations of microbes can immobilize a significant portion of the inorganic N by converting it to protein and stabilizing it into a non-leachable, non-volatile organic nitrogen. When those organisms die, they are decomposed by other microbes and the N is slowly mineralized back into plant food. However, in order for soil micro-organisms to accomplish this, they must have energy in the form of organic carbon.

A problem occurs when inorganic N is applied on a constant, excessive, and indiscriminate basis causing the organic carbon to be depleted beyond a healthy level for soil life (see chapters 2 & 5). Not only is no organic carbon being added to the soil, but the decomposition of existing organic matter is being accelerated. If heavy applications continue, less N will be stabilized by the dwindling populations of bacteria and the efficiency of nitrogen use drops rapidly. Consequences of this scenario can also include: 1. Ground water pollution, 2. Insect problems, 3. Greater potential for disease, 4. Soil compaction, 5. Thatch buildup (in turf), 6. Decreased drought tolerance, and 7. A colossal waste of money from both the nitrogen lost and the synthetic controls required to combat all these new problems.

Figure 6-3 shows the reaction of two different families of soil bacteria (i.e. nitrogen fixers and decomposers) to the introduction of inorganic N fertilizer. The nitrogen fixing bacteria are, in this case, symbiotic to the roots of alfalfa plants. However, non-symbiotic nitrogen fixers react in the same way. The presence of inorganic N makes it unnecessary for these organisms to fix atmospheric nitrogen. They have essentially gone on the dole (a British term for welfare). The decomposers, on the other hand, are often stimulated into a feeding frenzy resulting in the release of organic N and the recycling of carbon back to the atmosphere in the form of CO_2 at a significantly faster pace. This activity occurs from the dismantling of the soil organic matter by these microorganisms. If one understands the value of soil organic matter, then this also comes at an expense.

INFLUENCE OF INORGANIC N
ON NATURAL N SYSTEMS
(ALFALFA PLANTS)

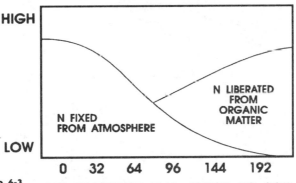

Figure 6-3
Waksman 1936

In 1950, a ton of inorganic fertilizer would, on average, boost yields of grain by forty six tons. By the early 1980's, the gain from that same ton of fertilizer was only thirteen tons of grain. The response difference is due largely to the depleted level of energy that is not being replaced by the inorganic inputs.

A solution to this problem is to use natural organic nitrogen whenever possible or small amounts of inorganic nitrogen mixed with sufficient quantities of OM such as compost, green manures or other sources of organic carbon. Natural organic nitrogen contains organic Carbon which can replenish the soil's energy reserves. Carbon is an essential component in sustaining the cyclical nature of the soil system and can help to balance the effect inorganic nitrogen has on the soil.

Figure 6-4 illustrates the natural cycle of nitrogen in the soil. It is important to note that this cycle would be physically impossible if carbon were not combined with nitrogen at the "Green Plants" stage. The carbon is essential for all the other functions in the cycle to occur.

Stable compost, with a C:N ratio of approximately 15-20:1, has much to offer in terms of soil conditioning, including large populations of beneficial bacteria, essential nutrients, and plenty of carbon. However, its Nitrogen analysis at one to three percent would mean

NITROGEN CYCLE

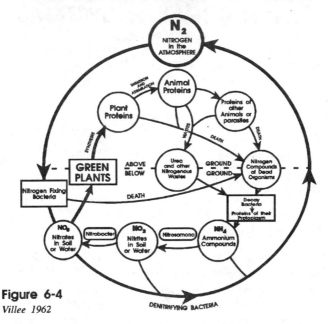

Figure 6-4
Villee 1962

applications of 1,500 - 4,400 lbs. to receive forty four lbs. N per acre (one lb N/1,000 sq.ft.). Compost is usually affordable, unless it has to be trucked over long distances, and it can be spread through most top dressing machines. Researchers at Cornell University and elsewhere have recently found that well aged compost (i.e. > two years) can also suppress many turf diseases (see chapter 4).

Products that contain a higher C:N ratio than 25-30:1 are probably not appropriate nitrogen sources. If the C:N ratio is too high, the nitrogen content is not sufficient to sustain the growing populations of decomposition bacteria. In addition, other sources of soil nitrogen are temporarily depleted.

Meals ground from beans or seeds usually contain a C:N ratio of approximately 7:1 but if shells are mixed in, the carbon value can get higher. Animal by-products, such as feather meal, blood meal or leather meal contain somewhere between three and eight parts carbon to one part nitrogen.

The C:N ratio of raw manures varies considerably depending on the animal it is derived from, that animal's diet, and the type and amount of bedding that is mixed with it. Raw manures are an impractical source of nutrient for many crops. They are difficult to apply, are aesthetically displeasing, can offend one's olfactory senses, are often replete with weed seed, and contain nitrogen that is unstable. Composting manures prior to application, rather than using them raw, is a very practical solution.

Many companies are claiming that their products contain organic nitrogen; however, they are deriving it from urea based ingredients. Urea is synthetic organic nitrogen with a C:N ratio of 0.4:1, and offers very little carbon to the biological activity in the soil. It has just enough carbon to call it organic.

Some companies mix inorganic nitrogen into fertilizer blends along with organic carbon from other sources. This combination can increase nitrogen efficiency if enough organic carbon is added. However, there is no set formula that determines the proper ratio. The C:N ratio in a natural soil system averages around 12:1. Natural organic nitrogen from plant and animal proteins ranges from three to eight parts carbon to one part nitrogen. No research was found that could quantify the ideal C:N ratio of fertilizers. Common sense dictates that C:N ratios somewhere between 12:1 (soil) and 3:1 (high N animal protein) would be appropriate. Heavy use of a fertilizer with an overall C:N ratio of less than 2:1 could eventually lead to soil organic matter depletion.

Sources of organic nitrogen vary in C:N ratios, and unfortunately, C:N information is usually not available on bagged, natural organic fertilizers. The only way to find out the C:N ratio of organic fertilizer is to either have it tested or to ask the manufacturer to do so. Some companies may already know the C:N ratio of their products. Table 6-2 indicates the C:N ratio of various ingredients. Feeding nitrogen into the soil is absolutely necessary at times, but applying it without carbon can eventually cause more problems than it is worth.

Natural organic sources of nitrogen are derived from proteins in plant and animal residues. The N in animal manures is utilized in large quantities by farmers for crops and by manufacturers who compost or ferment and bag it for sale to consumers at garden centers.

table 6-2	
INGREDIENT	**C:N RATIO**
Blood meal	4:1
Feather meal	4:1
Leather meal	4:1
Vegetable Protein meal	7-8:1
Animal tankage	3-5:1
Dried whey	7:1
Urea	0.4:1
Cow manure	11-30:1
Horse manure	22-50:1
Hen manure	3-10:1
Sheep manure	13-20:1

Composting and fermentation chemically change the manure into a more stable form of fertilizer, but it is accomplished by natural micro-organisms and is still considered a natural organic product. Composted manures are an excellent amendment to soils because of the high percentage of organic matter they contain, but they must be used in larger quantities because of their relatively low nitrogen content. To small scale gardeners, who have a tendency to overdose their soil with fertilizers anyway, this is a preferred material.

Some manure products are being produced by a new process that incorporates partial composting and quick drying. The result is a dry, granular manure that has a higher nitrogen content. However, this process of *semi-composting* will not stabilize all the N into a non-volatile, organic form, and strong odors are often painfully evident once the bag is opened.

Other animal residues that contain nitrogen are blood meal, feather meal, hoof and horn meal, meat and bone meal and leather meal. Some dairy by-products such as dried whey also contain some nitrogen. These products differ in nitrogen content but are all much higher than composted manures. Feather and blood, for example, contain approximately 12% N, but blood will release it much faster than feathers. The other animal by-products listed above contain between 5% and 10% N and release it at varying rates, depending mostly on how finely they are ground.

Residues from plants that are high in protein are also useful as sources of nitrogen. Meals ground from beans or seeds such as peanut, cocoa or soy contain between 3% and 7% N. However, most of these products are primarily used in animal feeds and priced as protein, which can make them prohibitively expensive as fertilizer ingredients. Sometimes the vegetable protein meals used in organic fertilizers have been rejected for feed because of aflatoxins (natural toxins that are a danger to livestock but innocuous in the soil). This rejection makes the material more economically practical to use as a fertilizer.

The only source of natural inorganic (mineral) nitrogen is a deposit of sodium nitrate that occurs in the Atacama Desert, located in Northern Chile. Chilean Nitrate (so called) is mined and purified through a physical process and is shipped around the world. This natural nitrate of soda contains 16% soluble N. Unlike many inorganic salts, Chilean Nitrate does not acidify the soil. It actually has a slightly neutralizing effect on soil pH.

As mentioned earlier, the difference between synthetic inorganic and natural organic nitrogen is carbon. Carbon is the fuel from which all living organisms derive their energy. The carbon cycle illustration (figure 6-2) shows how this essential element is taken out of the air by plants and utilized by every living thing on earth before being returned to the atmosphere as carbon dioxide. Also illustrated is the nitrogen cycle (figure 6-4). It is important to note that at the "GREEN PLANTS" stage nitrogen is combined with carbon to form proteins. This carbon is necessary for all the other functions in the cycle to occur.

SOURCES OF NITROGEN (N)

Blood meal contains approximately twelve percent N in a form that is broken down easily by soil organisms and made available to plants in a relatively short period of time. It is best utilized when lightly incorporated into the surface of the soil. Unfortunately, this ingredient is as expensive as it is good. Dried blood is used extensively as a feed ingredient because of its very high protein content (approximately 80% crude protein). Unfortunately, this use makes it scarce and costly as a fertilizer. Competing with the feed market makes many sources of natural organic nitrogen untouchable.

Feather meal, like blood, assays to about 12% nitrogen, but it is derived from a very different form of protein. Keratin, a protein that occurs in hair, hoofs, horns and feathers, is very indigestible when fed to animals as protein or introduced to the soil as fertilizer. The structure of keratin is very tight and not easily broken down by soil bacteria. This attribute makes feathers an excellent long term source of nitrogen but not appropriate for the plant's immediate needs. Many companies provide preliminary hydrolysis (i.e. decomposition) of feathers by autoclaving, a process that cooks the feathers with steam and pressure. This step requires energy and raises the cost, while providing a disproportionate improvement in nitrogen availability. The microorganisms that seem to be most adept at degrading feathers are often found in the bird's manure. Therefore, a mixture of feathers and manure (from the same birds) will improve availability without increasing costs significantly. This practice would only be appropriate for farmers situated on or near feather producing operations. As processes are developed that make feathers a more available nitrogen source, the costs will continue to rise; not only from the expense of the process, but also because the product becomes more attractive to the feed market as a digestible source of protein.

Leather meal is an excellent source of organic nitrogen but, unfortunately, is laced with controversy over chromium, an element used in the tanning process. Leather meal that comes to fertilizer companies is usually the residues from the planer, a machine that evens the thickness of the hide. The planer shavings contain between 10 - 12% N and sometimes as much as 2.5% chromium. Chromium is used to stabilize the leather (i.e. increase its resistance to decomposition) for a longer useful life. Leather manufacturers used to use aluminum but found that chromium lasted longer. The reaction of both aluminum and chromium in the soil are almost identical. In a pH that is appropriate for most crops both elements become unavailable to plants.

Leather and chromium have been studied for more than forty years by researchers of universities, the USDA, and the leather industry. The general consensus amongst the academics is that it is innocuous to the soil environment when used as a fertilizer. Of course, some disagree. Most of the studies have shown that even under conditions where chromium is toxic to plants, it is still well beneath the toxic level for animals. In fact, some studies indicate that it is beneath the minimum daily nutritional requirement of most animals, including humans. Leather is generally not used as a feed but is allowed as an protein supplement for hogs.

Vegetable Protein meals (VPM) (e.g. cocoa meal, soybean meal, cottonseed meal, peanut meal, etc.) are usually lower in N that animal by-products but higher in carbon (C), an essential component of the soil system. Many VPM are used in animal feeds which can increase the cost beyond a reasonable level for fertilizers. However, many are waste components of the feed or food industry making them attractive ingredients for natural fertilizers. Some VPM are tested regularly for aflatoxins (i.e. a natural toxin produced by fungi) before being used in feed rations. These toxins can be lethal to animals but are innocuous in the soil. A positive test of as little as two parts per billion can send a one hundred ton rail car load to the landfill. Fortunately, those rejected VPM can be utilized as valuable components of natural fertilizers. Many, such as soy meal, cotton-seed meal and peanut meal contain between 6 - 8% N along with some phosphorus (P) and potassium (K). Ground cocoa shells and coffee wastes are typically around 2 - 4% N. VPM are excellent ingredients to natural fertilizers because of their consistency, nutrient value and organic carbon content.

Animal Tankage (AT) is the dried, ground remains of a slaughterhouse operation and contains mostly meat and bone meal. An analysis of 9-4-0 is not uncommon with this ingredient. AT is also used in the feed industry for protein and phosphorus. Aside from the high cost, there is also the problem of putrefaction. When meat begins to break down biologically, it can produce strong odors, especially if it is not turned into the soil. These smells can be offensive and attract animals that may cause damage to crops.

Dried whey is the dehydrated remains of the cheese making industry. Dried whey is usually very expensive because of its value as a feed ingredient, but some whey processing companies are drying the residues from their cleaning operation. These residues usually contain some acids used to separate the whey from pipes and tanks, making it unsuitable for animal nutrition and less expensive for fertilizer use. Unfortunately, very few companies are doing this and the availability of this waste resource is limited. However, as organic waste disposal regulations become more strict, more of this product may become available. With a typical analysis of 5-9-1, whey is also an excellent source of phosphorus.

Fish meal has a typical analysis of 10-0-0 but is not as good as it looks. Again, competition from the feed industry makes fish meal prohibitively expensive to use as a fertilizer. Odors from this product can cause real problems with animal pests, not to mention hired help. The consistency of the meal is oftentimes too dusty to apply through conventional equipment. Fish emulsion, the soluble portion of the protein from fish waste, is often used in agriculture or horticulture but commonly, only as a supplement in a soil fertility program. Actual soil improvement is rarely accomplished with strictly liquid feeding programs because very little organic matter is being applied. Odors are often a problem with the emulsion too. Some companies have successfully added odor masks or neutralizers such as citrus extract to mitigate the problem.

Natural Nitrate of soda (NNS) is not an organic source of nitrogen. However, it is natural and is allowed in moderate quantities by many organic certification groups. Natural nitrate of soda (NNS), also known as Chilean Nitrate, is a mined product from a desert in Northern Chile. It is the only known deposit of this mineral salt in the world. NNS has an analysis of 16-0-0 and is considered soluble. NNS is commonly used in New England and elsewhere on organic farms as a form of nitrogen that is available to plants in cold soils. The microbial activity needed to mineralize natural organic nitrogen (protein) is suppressed during those times of the year when the soil is cold.

The sodium (Na) content of sodic soils makes NNS incompatible in arid and semi arid regions. It contains 26% Na. Na, in small quantities, does not cause damage to the soil ecosystem. Plants can utilize this element, and it is an essential nutrient for most animals and other organisms. The nitrate is utilized directly by the plant and does not need to be biologically processed. However, microorganisms will also utilize this source of nitrogen for the production of protein and amino acids. Applying NNS along with an organic amendment such as cocoa meal, peanut meal or compost will increase the efficiency of both products. NNS should not be relied upon as a sole source of N.

Urea is a synthetic organic source of N made from natural gas. Although urea contains some carbon, it is not enough to provide any significant energy to the soil. Unless urea is coated or processed in some way to slow its release into the environment or immediately incorporated into the soil, its N can volatilize (i.e. be lost into the atmosphere). Urea contains 46% N. Like nitrate of soda, urea should be used along with an organic amendment to increase its efficiency. Groups that certify farm produce or landscape services as organic prohibit the use of urea.

PHOSPHORUS

The occurrence of phosphorus (P) in any form can be traced back to mineral deposits of phosphate rock (apatite). Bone meal contains P that the animal derived from plants, and those plants probably utilized P from some form of rock phosphate. Neither bone meal nor rock phosphate are considered organic because they do not contain carbon. The leftover protein residues in raw bone meal (not removed by steam or precipitation processing) are organic, but raw bone meal is not used by many manufacturers because of strong odors that can offend the user and attract animal pests into gardens. Organic phosphorus occurs in manures and in plant residues but usually in relatively small quantities because plants utilize less phosphorus during their growth period compared to their need for other macro nutrients. The exceptions are dairy by-products that contain a concentrated amount of available phosphate. Some of these by-products are organic (i.e. contain carbon) and some are not. Organic gardeners or land care specialists who need to correct P deficiencies usually look to natural mineral (inorganic) sources such as bone meal or rock phosphate, rather than to natural organic.

Conventional phosphate fertilizer is made by reacting phosphoric acid with rock phosphate to form triple super phosphate. Another popular formula is to combine ammonia with phosphoric acid to form either MAP (monoammonium phosphate) or DAP (diammonium phosphate). All three forms are considered, by organic growers, to be highly acidifying to soils, and cause pH related problems. In addition, a high percentage of their total phosphate content is in an available form, making them susceptible to:

1) Fixation, because available phosphate bonds easily to many different soil constituents making them and itself unavailable and,

2) Causing pollution, because surface run off will carry dissolved phosphate into rivers and streams causing eutrophication.

SOURCES OF PHOSPHORUS

Rock Phosphate (RP) is a mined raw material used in the production of most of the refined phosphate products on earth. Huge deposits in the southern United States and elsewhere in the world are thought to have been formed by the accumulation of fossil shells on the ocean floor (over a period of millions of years) and brought to the surface by some broad tectonic event. Phosphate rock is also known as the mineral apatite, a chemically impure tri-calcium phosphate. RP is mostly insoluble. Only three of the 30% total phosphate content is considered readily available. Because of its insoluble nature, it has to be finely ground to be of any use in agriculture or horticulture. However, this process makes it dusty and difficult to work with. Applications of RP are usually much heavier than its chemically refined cousins but can last five years or more before another application is needed. Although its release in the soil is slow, it is often more efficient. The level of available phosphate from RP is largely dependent upon microbial activity, which typically coincides with the season of plant growth. In conventional fertilizers, a highly available phosphate is subject to losses from surface runoff or chemical fixation in the soil. It is thought that RP also stimulates the growth of populations of beneficial microbes, increasing the soil system's efficiency. Aside from its phosphate content, RP also contains 33% calcium and has one-fifth the neutralizing value of lime. Because RP is a natural raw material, it also contains some trace minerals.

Black Rock phosphate (BRP) like RP is an apatite mineral found in the Carolinas. Its consistency is more course, like sand, and it is easier to handle. Black rock gets its name and color from the small amount of (inorganic) carbon it contains. Laboratory tests indicate that BRP has a higher amount of available phosphate (from 4% to 8% depending on the lab) than finely ground rock phosphate. Like RP, BRP contains calcium and a list of trace elements.

Colloidal Rock phosphate (CRP) (also called Soft Rock Phosphate) is a waste product from the phosphate mining industry. Rock phosphate is first washed before it is shipped to refineries. The residues from the procedure are stored in large lagoons and are dried by the sun. A material is left that contains 18% phosphate (2% available), 21% calcium and a plethora of trace minerals. Unlike RP and BRP, CRP contains colloidal clay, a very fine clay particle that can help bind sandy soils enabling them to retain more water and nutrients. Because of the fineness of these particles, CRP can be dusty and difficult to work with.

Calcined Rock Phosphate is rock phosphate that has been heated. The process weakens the chemical bond between calcium and phosphorus and creates a product with a higher amount of available phosphate. Phosphate is usually calcined for a specific use and is rarely available as a waste by-product.

Soap phosphate (also called organophos) is a calcined rock phosphate that the soap industry used as a whitener in detergents. However, when it became unlawful to sell detergents with phosphates in certain states, soap companies were forced to remove it from their products. The recovered waste was then used as a fertilizer ingredient. Very few soap manufacturers are using phosphates as cleaning agents now; consequently, soap phosphate is scarce.

Raw bone meal comes from raw, uncooked bones that are ground and packaged. The protein residues from meat, cartilage and marrow contribute from 5% to 7% N to this amendment. The phosphate content typically ranges from 12 - 18%. Raw bone meal is also used in the feed industry and can be expensive. Some of the problems associated with using raw bone meal are the odors from the putrefying protein, the pests attracted by the smell, and the increase in labor from a moist and caked consistency that is difficult to spread.

Steamed bone meal is raw bone that has been put through a process similar to the steam cleaning of surgical tools. Because this step removes much of the protein residue, it produces a bone meal with lower nitrogen, fewer odors, and one that is easier to handle. The steaming process was originally developed to kill pathogens harmful to livestock who were being fed the bone meal. Steamed bone meal is expensive because of the energy needed for the process and the competition from the feed industry.

Precipitated bone meal (PBM) is a by-product of the gelatin industry. Certain protein residues attached to bones are necessary in the manufacture of gelatin. Part of the process requires the separation of phosphate from the other compounds. To do this, lime is mixed with bones dissolved in an acid solution. The calcium from the lime reacts with phosphorous from the bones to create a calcium phosphate compound. This compound precipitates out of solution and is dried to be used as an ingredient in feed and fertilizer. PBM contains approximately 45% phosphate and, because of its flour-like consistency, most of that is available. Unfortunately, the fine particles make the product dusty and difficult to work with. Some fertilizer companies are pelletizing it so that it will flow through conventional equipment.

There is no nitrogen in precipitated bone meal. PBM is used in the feed industry as a supplement providing calcium and phosphate but not as much as a product containing protein. The product is relatively expensive but, because of its high phosphate content, calculates into much less per pound of P than any of the other bone meals.

Whey meal (see Dried Whey in the nitrogen section.)

Precipitated milk Phosphate is made by essentially the same process that produces precipitated bone meal. Manufacturers that are extracting lactose from milk or whey react lime with the dissolved phosphate to precipitate out a calcium phosphate compound. The product contains approximately 23% total phosphate, of which 19-20% is available. Like precipitated bone meal, milk phosphate is dusty and difficult to work with. If the product were to be pelletized or granulated, it would be an ideal soil amendment.

Super phosphate is made by reacting sulfuric acid with rock phosphate. The resulting compounds are an available phosphate and gypsum (calcium sulfate). This process creates a product with an analysis of 0-20-0 with 20% calcium and 12% sulfur. It is neutral in its reaction with the soil (i.e. it does not alter soil pH).

Unfortunately, this product is prohibited by the various committees that certify food as organically grown. This is somewhat ironic because the same process occurs in nature when sulfur, oxidized by bacteria, combines with free hydrogen in the soil to form sulfuric acid. When the acid reacts with apatite rock, super phosphate is produced. The concern of the various certification committees is that if too much available phosphate is applied at once, surface run-off can carry away some of the phosphate, causing pollution.

Triple super phosphate is made by reacting rock phosphate with phosphoric acid. Like Super Phosphate, Triple Super is also prohibited for use by certified organic practitioners.

POTASSIUM

Potassium, like phosphorus, is a mineral in its original form. When used by organisms, it is changed into an organic form and occurs in the fluid portion of cells. If a plant is burned, the remaining ash would contain much of the potassium (and other minerals) that it utilized during growth. However, the carbon content would be released as CO_2 and other gases, and the remaining nutrient is no longer organic. Natural sources of potassium (K) include rock dusts, greensand, and potassium salts, such as sulfate of potash or sulfate of potash, magnesia. The ashes of various seeds' hulls, such as sunflower, barley or buckwheat are also rich in potassium. Compost usually contains at least 1% potash (K_2O) depending on what it is made from. This may appear to be an insignificant amount until you consider how much compost most gardeners apply to their soil. An application of compost with 1% potash, that is spread approximately 1/4 inch thick contains nine lbs K_2O / 1000 sq. ft.

Muriate of potash (potassium chloride) is a natural potassium salt but is prohibited by the organic community because of the chloride content, which they feel can acidify soils and inhibit the activity of many beneficial micro-organisms.

SOURCES OF POTASSIUM

Ash The term potash originally came from a old process where hot water was mixed with ashes and then filtered through sackcloth. The resulting solution, rich in dissolved potassium salts, was then dried and used as fertilizer. Farmers from that era may not have realized that there was more than just potash dissolved in the liquid. Ashes constitute most of the mineral accumulated by the plant over the course of its lifetime. These may include magnesium, calcium and trace minerals, depending on the type of plant and the environment it was growing in. When using ash in agriculture, it is a good idea to have it analyzed by a lab to see what one is actually applying. Sometimes applications of ash will create excessive levels of one nutrient while correcting the deficiency of another. Companies that broker ash for agriculture or horticulture use usually have it tested on a regular basis.

Greensand is a natural iron potassium silicate mineral also known as glauconite. It has the consistency of sand but is able to absorb ten times more moisture, making it an exceptional soil conditioner. Greensand contains potassium, iron, magnesium, calcium and phosphorus plus as many as thirty other trace minerals.

Jersey greensand, so-called from its only known place of origin...New Jersey, was deposited millions of years ago when the Garden State was still under water. It is mined primarily for water purification purposes, but more and more people in the agriculture business are demanding it for the soil. Benefits from greensand are, for the most part, unexplainable. If one brought some into an agriculture science laboratory and asked for an analysis, he would most likely tell you the product is worthless. However, numerous greenhouse trials show that there is much more to it than what someone might read on a lab report. Organic growers have, for years, extolled the virtues of greensand without really knowing how or why it has improved their crops. One possible explanation is the mineral content. The introduction of natural minerals has shown to improve soil by increasing populations of microorganisms. Greensand is an insoluble source of potash and trace elements, and releases them slowly. If there is an immediate need for available potash, it is suggested that a more soluble form of potash be applied in conjunction with greensand.

Sulfate of Potash, also known as potassium sulfate, is a naturally occurring potassium salt that contains 50 - 52% potash and 17% sulfur. Although sulfate of potash is soluble, it is accepted by most organic certification groups as a means to add potash to the soil. Soluble potash should be applied in accordance with the soil's capacity to hold cation nutrients (see CEC in chapter 7)

Sulfate of Potash, Magnesia commonly referred to as Sul-po-mag or K-Mag (brand names), is also a natural mineral salt and typically contains 22% potash, 11% magnesium and 17% sulfur. Like sulfate of potash, Sul-po-mag is soluble but is an acceptable material to use in the production of certified organic crops and the practice of organic land care.

Muriate of Potash, or potassium chloride, is a naturally occurring potassium salt but is prohibited by certification groups. It is considered detrimental to the soil because of the chloride content, which they feel inhibits the growth and establishment of beneficial bacteria and lowers the pH of the soil. A factor to consider is that the acidity of Muriate is equivalent to sixty seven pounds of lime per one hundred pounds of the material. As the pH of a soil becomes more acidic, populations of bacteria have a harder time surviving. However, experiments have shown that a soil environment with potassium from muriate of potash harbors far greater populations of microorganisms than an environment that is deficient in potassium.

TRACE ELEMENTS

Raw ingredients used to make natural fertilizers are inherently rich in trace nutrients. Mother nature likes diversity in the materials she creates. Humans, on the other hand, like to purify everything. Unfortunately, in doing so we tend to overlook the importance of those "contaminants" that we exclude.

Trace elements are essential for plants of any kind. However, the subtle differences between not enough and too much can easily injure or kill plant organisms. In most cases mineral soils provide ample amounts of trace elements to plants. Often, when trace elements are deficient, the cause is other factors such as incorrect pH, soil atmosphere imbalances, nutrient imbalances, or inadequate weathering mechanisms in the soil. Mono-cropping can also deplete the soil of certain trace elements because of the plant variety's constant demand for a specific nutrient.

Commercial preparations of trace elements are usually in the form of mineral salts or synthetic chelates. Because plants are very sensitive to excess trace elements, applications of these materials should be in accordance with an accurate soil test. Natural fertilizers made from raw, unrefined, materials usually contain a variety of trace elements which are released at a rate relative to the level of biological activity in the soil. Kelp meal or seaweed extract are also good sources of trace elements.

Plants utilize trace elements in very small quantities for the formation of enzymes and other organic components. Once combined with organic compounds they are called chelates. The more organic matter content in a soil, the richer it will be in chelates. Organic chelates are a major source of available micro-nutrients in the soil.

Natural sources of trace elements are both mineral and in the form of organic chelates. Most raw minerals such as rock phosphate, greensand, granite dust and basalt, are rich in trace elements but they are insoluble and rely on biological activity to make them available to plants. Organic sources include composts, manures and green manures, but the trace mineral content of these sources will vary depending on how much of these elements were in the environment where these sources were produced.

When addressing trace element deficiencies, natural sources may not be appropriate to use. The diversity of elements in natural materials may be fine for maintaining a balanced level in the soil, but these materials do not have enough of any one element to address a specific deficiency. For this reason, most organic certification groups allow synthetic, inorganic sources of trace elements to be used on a restricted basis. The use of concentrated commercial preparations of trace elements is not allowed without justification from a soil test.

Some manufacturers produce synthetic chelates containing specific trace elements which are in a highly available form. Others produce salts or oxides that contain trace elements, also in an available form. Great care must be taken when using these products because there is an extremely fine line between not enough and toxic levels of trace elements.

OTHER AMENDMENTS

Manure has been experimented with as a means of renewing soil fertility for hundreds of years. It is natural, it contains significant levels of organic carbon, it provides both soluble and insoluble nitrogen, along with most other essential plant nutrients; but, suprisingly, it is unacceptable to use in a certified organic program except in a few circumstances. Raw, uncomposted manure can cause some problems. Its nitrogen, for example, can be volatile (i.e. much of it can escape into the atmosphere). It is estimated that half of the nitrogen from topically applied manure is lost either by volatilization or leaching. Another problem with raw manure is the content of undigested seed from hay or pasture. The use of manure that contains significant levels of viable seed on row crops can be devastating and fosters a need for herbicide. Another concern of the organic certification groups is that manure used late in the season increases (to an unhealthy level) the amount of unprocessed nitrates and nitrites in produce. Since the main objective of organic certification is to insure safe and nutritious food, and to improve the soil environment, the concerns about raw manure are valid.

Compost used to be referred to as artificial manure back in the early part of this century. If made correctly, it is the other extreme to raw manure. Compost provides a stable source of nutrients and organic matter with no weed production and no safety or environmental controversy. It is the quintessential fertilizer. In situations where it is the only available source of plant nutrients, the fertility and tilth of the soil improve immeasurably. Many crop problems associated with environmental stress are mitigated or disappear when compost is used. The nitrogen in compost has, through a biological process, been converted into protein which will not leach or volatilize into the atmosphere. The mineral content is organically bound, for the most part, making it easier for organisms to access it. The organic matter from compost adds significantly to the water and nutrient holding capacity of the soil. The drawbacks to compost are in the shipping and handling of the product. Its freight sensitive nature requires a relatively local source and, other than a topdressing machine or manure spreader, the only way to apply compost is by hand.

Green manures are crops, usually legumes, that are grown specifically to improve the soil's structure and fertility. It is important not to confuse green manures with cover crops that are grown for a different short term purpose. Cover crops are planted to hold soil, retain nutrients and control weeds in between annual crops. Green manures are planted, usually for a period of one year or more, to replenish the soil's reserve of nutrients through the fixation of nitrogen from the atmosphere and the scavenging of mineral from soil resources. Experiments have shown that some legumes, such as hairy vetch, can produce a significant amount of nitrogen on a short term, cover crop basis. However, most green manures require a longer period of time to produce the amount of organic matter lost to row crop cultivation.

Kelpmeal is a source of naturally chelated trace elements that can increase the health of both the soil and the plants. Unfortunately, the product is very expensive to use as a soil amendment unless it is locally available (i.e. found on coastal shores) and can be easily dried and ground. The more practical method of utilizing kelpmeal is to use it as an animal feed supplement. This improves production from the animal and enriches the manure for use in compost or on crops.

Seaweed extract is produced by dissolving many of the chelated trace elements and growth hormones from the seaweed plant and, either bottling it as a liquid or drying it into a soluble powder. Seaweed extract is used for a variety of purposes including improved seed germination, disease and insect resistance, inhibiting senescence, stimulating root growth, and providing essential chelated micro-nutrients that are needed for the production of vital enzymes. The growth hormones in seaweed extract have been shown to significantly increase a plant's resistance to stress. These hormones are produced naturally by most plants, but the production is inhibited by stress such as extremes in temperature, drought or reproduction. If external sources are available, the plant can weather environmental stress conditions that naturally occur and remain a stronger organism. The plants are then better able to resist problems such as disease or insect attack that weaker plants are more susceptible too. Seaweed extract does not usually carry a guaranteed N-P-K analysis on its label because it is not a fertilizer. It is more accurately classified as a bio-stimulant. Seaweed extract can be applied as a foliar or root feed but is always used as a liquid. It is a relatively inexpensive product and allowed by most certification groups.

Humates are naturally occurring components of humus mined from rich organic deposits such as old peat bogs. They are also classified as bio-stimulants. Humates improve the nutrient and water holding capacity of the soil. Their most significant value, however, is somewhat of a mystery. Research has shown that plants can utilize some organic compounds from humates and exhibit favorable growth responses. Humates occur naturally in a healthy soil. Some researchers have shown that added humates have little or no effect in soils that are already rich in organic matter. Others dispute those findings, showing favorable responses in a variety of conditions. Common sense dictates that there is a maximum response plants can exhibit.

Beneficial Bacteria are used as an inoculant by many companies that make natural fertilizers, bio-stimulants and compost starters. There is evidence that the use of these organisms has an enhancing effect but many experts argue that the natural indigenous bacteria will provide as good, if not better, inoculation if given the fuel needed for growth. The bacteria used in commercial preparations are usually substrate specific, meaning that they are adept at breaking down specific components of organic matter and will not compete for survival very well if that substrate is unavailable. Others argue that the competition from naturally occurring bacteria is too overwhelming for introduced varieties to be effective. Since there are probably no two environmental situations on earth that are identical, it would be difficult to identify instances where these organisms are beneficial and where they are not. The use of beneficial bacteria is, at best, an efficient method of stimulating necessary bio-activity or, in the worse case scenario, a waste of some money.

Ground Limestone (calcium/magnesium carbonate) is both a pH neutralizer and a source of calcium and/or magnesium. The carbonate component of lime is what neutralizes soil pH by chemically converting hydrogen ions into water and carbon dioxide. Calcium and magnesium are essential plant nutrients. Lime is a critical component of agriculture and horticulture but there is such a thing as too much. Soils with too much lime foster high populations of bacteria that can use up too much of the soil's energy at once accelerating the depletion of organic matter. A high pH can also limit the availability of many trace elements in the soil.

Aragonite is a calcium carbonate mineral like limestone that comes from sea shells such as oyster shell. It is used in lieu of lime in situations where the soil is already high in magnesium, and dolomite (high magnesium limestone) is the only liming material available. Aragonite has approximately 89% of the neutralizing value of lime.

Gypsum or calcium sulfate is a naturally mined product containing 23% calcium and 17% sulfur. Gypsum is used extensively where calcium is needed and a pH change is not desired. Because gypsum contains no carbonate or oxides it does not neutralize hydrogen. It is also used to alleviate compaction in certain clay soils where a chemical reaction with gypsum causes some granulation. Gypsum is sometimes used to de-salinate soils where road salt has caused damage. There is also some evidence that it can provide insect control for certain turf pests such as chinch bugs or grubs (university data is not yet available).

Epsom Salts, or magnesium sulfate, is used where magnesium deficiencies occur in already alkaline conditions. Magnesium oxide (mag-ox) is a more common and a less expensive source and because of its concentration (60% Mg) mag-ox has only a minimal effect on pH.

BLENDED FERTILIZERS

Soil amendments that offer only one or two primary or secondary macro-nutrients should only be used for correcting deficiencies of those nutrients or for balancing excessively high levels of other nutrients. Mixing natural amendments together as a blended, complete fertilizer is a practical method of offering the soil a broad selection of nutrients that both plants and soil organisms can benefit from and process as needed. It can also mitigate soil compaction and energy consumption by reducing the number of trips a tractor or other heavy equipment travels over a field. Companies that blend fertilizers can customize mixes to suit a particular soil condition or specific crop needs. However, if nutrients are introduced as raw materials, there is some natural control over what is made available by soil organisms and what is not.

THE FERTILIZER LABEL

In the U.S., companies that produce fertilizers for sale must comply with labeling laws. These laws are written and enforced by each state and, although uniform guidelines are recommended, the states sometimes differ in their labeling requirements. However, most states require that a fertilizer label contain the following information (see figure 6-5):

UNIFORM FERTILIZER LABEL

NET WEIGHT
BRAND
GRADE (N-P-K)
GUARANTEED ANALYSIS

Total Nitrogen (N)	?.?%
?.?% Water Insoluble Nitrogen	
?.?% Water Soluble Nitrogen	
?.?% Ammoniacal Nitrogen	
?.?% Nitrate Nitrogen	
?.?% Urea Nitrogen	
Avail. Phosphoric Acid as (P₂O₅)	?.?%
Soluble Potash (K₂O)	?.?%

SOURCE OF NUTRIENTS (DERIVED FROM:)
NAME AND ADDRESS OF MANUFACTURER

Figure 6-5

1. Net weight expressed in pounds and in Kilograms (optional).

2. Manufacturer's name and address.

3. Grade. This is a brief outline of the guaranteed analysis in the format: N-P-K. N represents the total nitrogen, P is the available phosphoric acid and K is for soluble potash.

4. Brand name that the manufacturer has given to the product.

5. Guaranteed analysis is the manufacturer's guarantee of the plant nutrients contained in the package or bulk load. The nitrogen value is always indicating the total amount of nitrogen. Most states require that the different types of nitrogen (e.g. water soluble and water insoluble, ammoniacal and nitrate) be described just below where the total nitrogen appears. The percentages in this description must add up to the same value given for total nitrogen. Phosphorus is always expressed as available phosphate (P_2O_5) and Potassium, as soluble potash (K_2O).

The guaranteed analysis can also include secondary macronutrients such as calcium, magnesium and sulfur or trace elements that are considered essential to plant growth.

Bio-stimulants, soil conditioners and long term (slow release) insoluble minerals are not allowed to be mentioned on the label.

Fertilizer labels always express the phosphorus content of the product in terms of Available phosphoric acid (P_2O_5). However, P_2O_5 is not phosphoric acid and it is not a form of phosphorus that plants use. Phosphoric acid is actually expressed as H_3PO_4. Plants utilize phosphorus as HPO_4, H_2PO_4 and, in some cases, H_3PO_4. The only conditions under which P_2O_5 exists is when phosphate is heated to 650° C. The same is true for K_2O (so-called *soluble potash*). Heating phosphate and potash is a common way of testing for their content. Consequently, plant food control officials require labeling in this manner.

6. The *Derived from* list is a list of materials used to obtain the guaranteed analysis. It does not include any soil conditioners, etc. that may be present in the blend. However, if the consumer wants to be discriminating about the kind of fertilizer ingredients he uses, this is an important section of the label to look at.

7. Some states require that the Potential acidity of the fertilizer be expressed on the label in terms of the amount of lime (calcium carbonate) it would take to neutralize it. A few states want to see the amount of chlorine in the fertilizer on the label.

The design of all the labeling regulations is to protect the consumer. Unfortunately, the more complex the label becomes, the less consumers will look at it; and the label has to be read if it is going to provide protection.

CAN ORGANIC FERTILIZERS COMPLETELY REPLACE CHEMICALS?

The answer is *YES*, it is done all the time, but the switch must be accompanied by certain changes in cultural practices. The key ingredient necessary to grow anything organically is information. You need to know why you have a problem as opposed to how to treat the symptom.

The professional landscapers or lawn care specialists will find it somewhat easier to be organic than the food, feed or fiber producers. This is because they are not harvesting nutrients from the soil that inevitably must be replaced. However, to do their job without synthetic chemicals requires an understanding of the natural soil/plant system. They need to know how the system encourages growth and, at the same time, how it discourages problems. They also need to understand the limitations of their environment. Land care specialists who have been organic for many years have discovered that if the soil system is fed and stimulated correctly, fewer inputs and less labor is required to grow better quality plants.

The farmer who grows organically understands the same principals. He also knows that rotations, green manures and composting are necessary to rebuild his soils depleted by harvests. Most experienced organic farmers do not purchase very much fertilizer. They make their own with on-farm resources. Third world farmers who can't afford commercial fertilizer have been doing this for thousands of years.

Unfortunately, what most people really want to know when they ask the question (i.e. can chemical fertilizers be completely replaced by organic fertilizers?) is not, "can I do what I'm doing organically?" They want to know if they can substitute organic fertilizers for conventional ones in a chemical feeding program without modifying their cultural practices.

Sure you can. But don't expect a perfect fit.

Sources:

AAFCO, 1990, Official Publication 1990. Association of American Feed Control Officials. Atlanta, GA

AAPFCO, 1990, Official Publication #43. Association of American Plant Food Control Officials. West Lafayette, IN

Albrecht, W.A. 1938, Loss of Organic matter and its restoration. U.S. Dept. of Agriculture Yearbook 1938, pp347-376

Bear, F.E. 1924, Soils and Fertilizers. John Wiley and Sons, Inc. New York, NY

Brady, N.C. 1974, The Nature and Properties of soils. MacMillan Publishing Co. Inc. New York, NY

Buchanan, M. and S.R. Gliessman 1991, How Compost Fertilization Affects Soil Nitrogen and Crop Yield. Biocycle, Dec. 1991. J.G. Press Emmaus, PA

Chu, P. 1993, Personal communication. A&L Eastern Agricultural Labratories. Richmond, VA

Lucas, R.E. and Vitosh, M.L. 1978, Soil Organic Matter Dynamics. Michigan State Univ. Research Report 32.91, Nov 1978. East Lansing, MI

Nosko, B.S. 1987, Change in the Humus for a Typical Chernozem caused by fertilization. Soviet Soil Science, 1987 v19 July/August p67-74

Novak, B. 1984, The Role of Soil Organisms in Humus Synthesis and Decomposition. Soil Biology and Conservation of the Biosphere. pp 319-332

NRAES, 1992, On Farm Composting Handbook. Northeast Regional Engineering Service #54, Cooperative Extension. Ithica, NY

Parnes, R. Fertile Soil: A growers Guide to Organic & Inorganic Fertilizers. Ag Access, Davis, CA

Senn, T.L. 1987, Seaweed and Plant Growth. No publisher noted. Department of Horticulture, Clemson University. Clemson, SC

Silkina, N.P. 1987, Effects of High Nitrogen Fertilizer Concentrations on Transformation of Soil Organic Matter. University of Moscow Soil Science Bulletin 1987, v42 (4): 41-46.

Singh, C.P. 1987, Preparation of High Grade Compost by an Enrichment Technique. I. Effect of Enrichment on Organic Mater Decomposition. Biological Agriculture and Horticulture 1987, vol 5 pp 41-49

Smith, G.E. 1942, Sanborn Field: Fifty Years of Field Experiments with Crop Rotations, Manures and Fertilizers. University of Missouri Bulletin #458. Columbia, MO

SSSA# 19. 1987, Soil Fertility and Organic Matter as Critical Components of Production Systems. Soil Science Society of America, Inc. Madison, WI

Villec, C.A. 1962, Biology. W. B. Saunders Company. Philadelphia, PA

Wallace, A., Wallace, G.A. and Jong, W.C. 1990. Soil Organic Matter and the Global Carbon Cycle. Journal of Plant Nutrition 1990 v13 (3/4): 459-456

Waksman, S.A. 1936, Humus. Williams and Wilkins, Inc. Baltimore, MD

White, W.C. and Collins, D.N. (Editors) 1982, The Fertilizer Handbook. The Fertilizer Institute. Washington, DC

Chapter 7
TESTING THE SOIL SYSTEM

As the world population expands into critical dimensions, the importance of judicious utilization of soil resources grows with it. Unfortunately, too few land care professionals are taking advantage of an available tool designed to dramatically increase the efficiency of agricultural or horticultural inputs. The tool is called a soil test. Soil tests are limited in terms of the scope of information they can provide. However, the basic notion of balanced fertility is the fuel for a functioning soil system.

Some professionals fail to understand that the soil is not just a growing medium. It is a biological system that functions in a symbiotic relationship with all plants. The best way to follow the changes that occur in this system from land use is with a soil test.

SAMPLING PROCEDURE

The soil is like an urban community in that no matter where we knock, someone different will answer the door. In the soil, it would be rare if two samples could be found, even if they were drawn a foot away from each other, that produced the exact same test results. So it is extremely important to get a good representation of the entire area being evaluated. The test results will only be as useful as the sample is accurate. Figure 7-1 shows an example of a sampling pattern usually recommended to insure results that are relative to the overall condition of the area. The number of samples taken should depend on the size of the area. The more samples taken, the better the representation.

ONE ACRE LOT

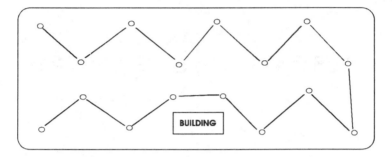

SUGGESTED SAMPLING PATTERN
Figure 7-1

Noticeably good or bad areas should be avoided (unless evaluation of a specific spot is needed). The conditions in these areas are extremes of one sort or another and will adulterate the average reading of an area. For obvious reasons, freshly fertilized or limed areas should also be avoided.

Very clean tools should be used for gathering samples. A small amount of rust on a shovel could be interpreted from a test as a good place to start an iron mining operation. An example of an incident that gave misleading results from contaminated tools is a client who used the same shovel to draw samples that he normally uses to clean out ashes from his wood stove. The results of the test were so far off that it may have well been from a lawn on the planet Potassium.

Soil sampling tubes are the most efficient and accurate tools for drawing soil even if sampling occurs only once a year. They significantly increase the speed of the sampling procedure while decreasing labor costs and the chances of contaminating the sample. These tubes are durable and relatively inexpensive. For sources see table 7-1.

Avoid wet or frozen samples. The proper consistency of a soil sample for analysis is moist but not soaked. The sample should ball up when squeezed, no water should drip out and the ball should crumble easily. Drying out the sample is an acceptable procedure if necessary, but to preserve the most original conditions of the soil, samples should not be drawn until the soil is at a proper moisture level. If the soil in question is naturally sandy and dry, do not attempt to moisten it.

The sampling depth for most applications should be about five to seven inches. Scrape off any surface debris such as roots or thatch from the top of the sample.

Table 7-1
SOURCES FOR SOIL SAMPLING TUBES
North Country Organics RR #1 Box 2232 Bradford, Vermont 05033 802/222-4277
Oakfield Apparatus Co. P.O. Box 65 Oakfield, WI 53065 414/583-4114
A.M. Leonard, Inc. 6665 Spiker Road P.O. Box 816 Piqua, OH 45356 800/543-8955
Nasco 901 Janesville Ave Fort Atkinson, WI 53538-0901 414/563-2446

When all the samples from a given area are drawn, mix them thoroughly and take a representative sample of approximately one cup to send to the lab. Again, to preserve accurate results, send the sample to the lab right away. Avoid letting it sit for an extended period of time.

CHOOSING A LAB

Soil test laboratories do not have nationwide standard procedures for testing soil. Regional methodologies may often be employed by most labs in a given area, but they can opt to follow any procedures they want. *Methods of Soil Analysis*, published by the American Society of Agronomy, is the most widely accepted manual of testing procedures, but the publication often describes many different, but valid, ways to test for nutrients. Consequently, if the same samples were analyzed by several different labs the results may come back looking as though the samples were taken from several different countries. Additionally, labs may not offer the same type of information as part of their standard test. The University of Vermont, for example, tests for aluminum to determine lime and phosphorus recommendations. A valid but relatively unique procedure.

On a functional level, the most important service a lab can offer is an accurate, reasonably priced test delivered on a timely basis. However, it is also important to determine the kind of information needed before looking at different labs. It is most cost effective if all the necessary information is offered on the lab's standard test as optional information can get expensive. Because of the variance in the results one can receive from different labs, it is a good idea to choose a reliable lab and stick with it, especially if one is comparing tests that represent the soil before and after a specific treatment.

Interpreting soil test results are not as difficult as they may seem. However, it takes a little time and practice. Unfortunately, many labs give mostly recommendations and insufficient test data needed for interpretation. Those labs cater to customers who only want the necessary information to make their soil balanced and fertile. Those customers are not really interested in learning how to interpret a lab analysis. However, the ability to interpret a soil test enables one to consider all the variables that can affect the performance of the soil in that specific area and to possibly save some money.

Most labs offer recommendations either automatically or as an option (extra cost) and are based on the data derived from the sample. Therefore, the lab recommendations are only as good as the samples taken. Recommendations are based on nutrient uptake of specific plants under average conditions. Normally, a lab will ask for more information such as type of crop, crop use, topography and previous treatments if they are to provide recommendations. Recommendations for customers who wish to implement organic practices are rarely available but more and more labs are beginning to respond to the need, so if organic recommendations are required, ask for them.

To determine nutrient balance and fertility in soil, look for a lab that offers the following information (listed in order of importance):

1. pH and Buffer pH
2. Percent Organic Matter
3. Cation Exchange Capacity (CEC)
4. Reserve and available phosphorus
5. Soil levels of exchangeable potassium, magnesium and calcium.
6. Base Saturation

pH

Many people believe that the initials pH stand for *potential hydrogen* or *power of hydrogen*. The H in pH does, in fact, stand for hydrogen but p is a mathematical expression which when multiplied by the concentration of hydrogen in the soil, gives a value between one and fourteen that expresses the acidity or alkalinity of the soil. Actually, soil pH values are rarely found to be under four or over ten. Values below seven are acidic and those above seven are alkaline. A value of seven is neutral. As values get farther away from neutral they indicate a stronger acid or base. A pH between six and seven is ideal for most plants.

Most labs offer pH in their analysis of your soil, but some do not mention how they are testing for it. Looking at the A & L sample analysis (see figure 7-2), notice that they give two pH values. The first is a water pH which determines the acidity or alkalinity of distilled water when mixed with an equal volume of the soil sample. This test only determines whether or not there is a need to lime or acidify the soil. The Buffer pH or pH_{SMP} test is done with a special solution that determines how much lime to apply. The buffer test is not used if the water pH is near neutral. If a lab only offers one pH value in its analysis that is not qualified as to how it was found, it is relatively safe to assume it is a water pH. In these cases, the lime recommendation offered by the lab must be relied upon because it is not offering enough information for anyone to determine an appropriate application rate. A few labs indicate salt pH on their analysis report. Salt pH is measured with consideration for seasonal variation of soil soluble salts that can cause changes in pH. Salt pH values are normally 0.5 to 0.6 below what a water pH test would show.

Figure 7-2

SAMPLE A&L
SOIL ANALYSIS REPORT

%OM	ENR	P1	P2	K	Mg	Ca	SOIL pH	BUFFER pH	CEC	% BASE SATURATION			
										K	Mg	Ca	H
4.5	104VH	83VH	121VH	139VH	187VH	640M	6.0	6.8	6.2	5.7	25.0	51.3	3.1
7.3	171VH	102VH	160VH	600VH	210VH	2800H	6.3	6.7	19.3	8.0	9.1	72.5	10.5

Figure 7-2 is a sample analysis from A & L Eastern Agricultural Laboratories in Richmond, VA.   %OM - percent organic matter, ENR - Estimated Nitrogen Release in pounds per acre (ppa), P1 - Available Phosphorus in parts per million (ppm), P2 - reserve phosphorus in ppm, K - Potassium in ppm, Mg - Magnesium in ppm, Ca - Calcium in ppm, Soil pH - Water pH test, Buffer pH - pHsmp, CEC - Cation Exchange Capacity.

ORGANIC MATTER

Most soils are called mineral soils because of the high level of rock mineral they contain. Soils were first formed on the earth by the weathering of rock into smaller and smaller particles by forces such as rain, frost, wind and erosion (See chapter 1). Eventually, rock particles became small enough to be utilized by organisms as mineral nutrient. The soil can be defined as that part of the earth's crust in which roots can grow. As photosynthesizing (autotrophic) organisms evolved, their needs consisted mostly of gaseous elements such as Oxygen, Nitrogen, Hydrogen and Carbon dioxide which were derived from the atmosphere. Their mineral needs where taken care of by the soil. As generations of these autotrophs cycled through life and death, decomposing organisms evolved and ultimately created humus (see chapter 2). After millions of years, as more and more organisms appeared, levels of organic matter increased, which increased the production of all organisms from the microbe to the plant and all the other life they supported.

Figure 7-3 shows the theoretical formation of different soil horizons over time. Note that the development of organic matter is not suggesting that the surface of the soil is 100% organic matter. It is simply the layer in which organic matter will accumulate. The process has a snow balling effect. However, organic matter has its own life cycle, and through oxidation, nitrification, and other

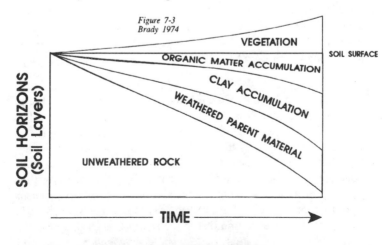

Figure 7-3
Brady 1974

SOIL DEVELOPMENT

SOIL COMPONENTS
Typical analysis of a well developed loam

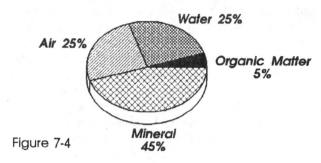

Air 25%

Water 25%

Organic Matter 5%

Figure 7-4

Mineral 45%

natural processes would, on average, not amount to more that five percent in the top six inches of a mineral soil (see figure 7-4).

Organic matter (OM) is a barometer of soil health. The population of organisms that is supported by soil organic matter is of immeasurable benefit to plants. More OM means more decomposers that recycle nutrients from plant and animal residues faster; more nitrogen fixing and mineralizing bacteria; more beneficial organisms that help dissolve mineral, translocate water from soil depths and help control pathogenic fungi; and more humus that increases the water and nutrient holding capacity of the soil. Humus acts like a sponge in the soil which expands and contracts as its moisture level changes. This activity within the soil increases porosity, which improves the movement of air and water. As all these organisms travel through their own life cycles, they create even more organic matter.

Burning is the most accurate method of testing for OM. What is left after combustion is ash, which is the mineral portion of the sample. The percent of OM can then be determined simply by subtracting the weight of the ash from the total weight of the dry sample before it was burned. A reading close to five percent in temperate regions is a good level. Readings significantly below that figure may be an indication that some sort of soil building program should be implemented. A look at the sample A & L soil analysis (figure 7-2), shows a column just to the right of % Organic Matter labeled ENR. This stands for *Estimated Nitrogen Release*, and indicates the amount of nitrogen, in pounds per acre, that can be released from the OM content throughout the season under ideal conditions. To convert this figure to pounds per 1,000 square feet, divide by 43.56 (because there are 43.56 thousand square feet in an acre).

Levels of organic matter naturally differ around the country for various reasons. The climate has a big influence on organic matter levels in different regions. Laboratories should be able to ascertain the average for areas that they service based on their own tests.

CATIONS, ANIONS AND EXCHANGE

To better understand cations (pronounced cat'-eye-ons) and anions (pronounced an'-eye-ons) a short review of elementary chemistry is in order. All materials that are on, under or above the planet's surface, whether liquid, solid or gaseous are made up of atoms of different elements that are bonded to each other. The force that holds these atoms together is electro-magnetic. It is the same force that holds dust to a television screen. However, in order for this magnetic force to work there must be both positive and negative charges present. Like charges do not attract.

There are ninety elements that occur naturally on earth. Most of these elements are made up of atoms that have either a positive or negative charge. Some, that are called inert, have no charge and rarely bond with anything. When the atom of one element bonds with the atom of another, the result is called a compound. The magnetic bonds between elements in many compounds satisfies the attraction each atom has for the other. However, in many other instances the attraction of one atom does not completely satisfy the magnetic force of the other leaving a net negative or net positive charge that is still available for another combination. These compounds are called ionic.

Ions are atoms of elements or molecules of compounds that carry either a negative or positive electric charge. The ions of elements such as hydrogen (H), calcium (Ca), magnesium (Mg) and potassium (K) have positive charges and are known as cations. Ions of phosphorus (P), nitrogen (N), and sulfur (S) have negative charges and are called anions. Some elements such as carbon (C) and silicon (Si) can act as anions or cations and bond to either charge.

Molecules of compounds such as nitrate (NO_3), sulfate (SO_4) and phosphate (PO_4) have negative (anionic) charges and can bond with cations such as H, Ca, or K.

In the soil, there are very small particles, called micelles (short for micro-cells) that carry an electro-negative or anionic charge. These particles are either mineral (clay) or organic (humus) and are referred to as soil colloids. Although small in comparison to other soil particles, colloidal particles are huge in relation to soil cations such as H, K, Mg or Ca. Cations are attracted to these colloids like dust is to a TV screen.

CLAY

Clay particles, if viewed through a powerful microscope, would appear as flat platelets adhering to each other like wet panes of glass (see figure 7-5). These particles are predominately comprised of silicon, aluminum and oxygen but, depending on the type of clay, can contain a plethora of different elements such as potassium (K), magnesium (Mg), iron (Fe), copper (Cu) and zinc (Zn).

CLAY PARTICLES
MICELLES

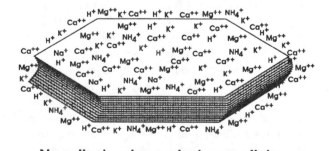

**Negatively charged clay particles
shown with typical plate-like
appearance and swarm of adsorbed cations**

Figure 7-5
White 1982

Clay's magnetism comes from the natural substitution of ions in its structure with other ions that don't completely satisfy the available negative charges. As more of these substitutions occur, the overall negative charge of the particle increases, and there is a relative increase in the amount of cations attracted to it (see figure 7-5). Different types of clays have different abilities for ionic substitutions, and their magnetic force differs accordingly. Clay particles that have greater substitution potential will create more magnetic force and attract and hold more cations. Some clay particles are not colloidal at all.

The surface area of clay is also important to consider. If one acre (six inches deep) of clay particles were to be separated and spread out, the surface area would be equivalent to the size of the State of Illinois times fifty. Since most of these ionic substitutions occur on the surface of the clay platelets, there is a relative increase in the soil's capacity to hold cations from the amount of exposed surface area of clay.

HUMUS

Humus particles are also colloidal in nature (i.e. they are negatively charged). Their cation exchange capacity (CEC) is even greater than clay per equal unit of weight but is influenced by soil pH (see figure 7-6).

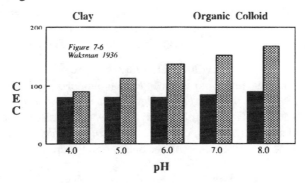

Humus gains its magnetic charge from the surface compounds that contain hydrogen. In soils with a near neutral pH, H is displaced from humus to participate in a number of different soil chemical reactions. When this occurs, a negative charge that previously held the H ion is left unsatisfied and available for another chemical bond. If base cations such as K, Mg or Ca are present in the soil solution

ORGANIC COLLOID

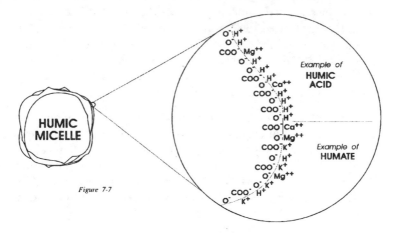

Figure 7-7

they will be attracted to the humic particle (see figure 7-7). Like clay, humus has a significant amount of surface area.

Clay and humus will oftentimes form colloidal complexes together, which not only enhances the overall cation holding capacity of a soil, but also changes the structure of clay soil into a better habitat for roots and organisms. Humus, through complicated chemical bonds, can surround clay particles and break up its cohesive nature that can prevent percolation of air and water through the soil.

Both clay and humus, with their unique structure and electrostatic (negative) charge, hold onto a tremendous amount of cation elements in such a way that the ions can be detached and absorbed by plant roots. Hydrogen (H^+) ions that are given off by roots are traded back to the colloids. This whole process, called *cation exchange*, is the major source of cation nutrients for plants (see Figure 7-8).

CATION EXCHANGE

The exchange of cations that takes place in the root zone is an essential component of the soil system. Minerals such as calcium, magnesium and potassium would be relatively rare and unavailable in topsoil if it were not for the colloidal nature of clay and humus.

Base cations are magnetically held by soil colloids. When root

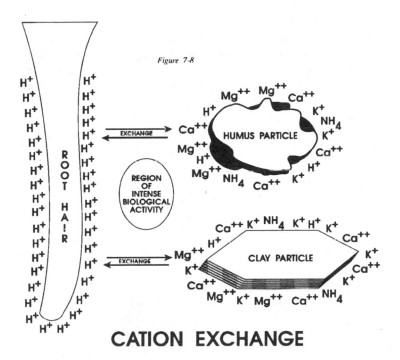

Figure 7-8

CATION EXCHANGE

hairs grow into proximity with colloidal exchange sites, those bases can displace hydrogen ions located on the root's surface and can then be absorbed by the plant. This exchange is essential, not only for the mineral needs of the plant, but also for the translocation of hydrogen throughout the soil (see figure 7-8).

HYDROGEN

Depending on soil pH and the level of exchangeable cation bases (i.e. Ca, Mg and K) in the soil solution, the hydrogen that is adsorbed to the soil colloid will eventually exchange again; however, this time with a base from the soil solution. Once the H^+ ion enters the soil solution a number of reactions can occur. It can combine with oxygen to form water, or with nitrate or sulfate ions to form strong inorganic acids (i.e. nitric and sulfuric acid). These acids can work to liberate more base cations from the mineral components in the soil. H^+ can also liberate and transform phosphate ions into phosphoric acid, which is the form of phosphorus that plants can utilize.

Soil carbonates react with H^+ ions to form carbon dioxide (CO_2) and water. This process is implemented when lime (calcium carbonate) is added to raise the pH of a soil. Hydrogen can also help form organic acids when reacting with residues of plants, animals and microorganisms. These acids can weather available minerals from rock particles in the soil (see chapter 1). This function is especially important in the zone surrounding the roots of plants called the rhizosphere. Within the rhizosphere, biological activity is very high and the pH is relatively low. The activity of H^+ ions in this zone releases mineral ions from soil particles that become available nutrients for the plant. Oxides, that are abundant components of many minerals, can also combine with H^+ to form water H_2O.

CEC

The cation exchange capacity of the soil is determined by the amount of clay and/or humus that is present. These two colloidal substances are essentially the cation warehouse or reservoir of the soil. Sandy soils with very little organic matter (OM) have a low CEC, but heavy clay soils with high levels of OM would have a much greater capacity to hold cations.

The disadvantages of a low CEC obviously include the limited availability of mineral nutrient to the plant and the soil's inefficient ability to hold applied nutrient. Plants can exhaust a fair amount of energy (that might otherwise have been used for growth, flowering, seed production or root development) scrounging the soil for mineral nutrients. Soluble mineral salts (e.g. potassium sulfate) applied in large doses to soil with a low CEC cannot be held efficiently because the cation warehouse or reservoir is too small.

Water also has a strong attraction to colloidal particles. All functions that are dependent on soil moisture are also limited in soils with low CEC. Organisms such as plants and microbes that depend upon each other's biological functions for survival are inhibited by the lack of water. Where there is little water in the soil, there is oftentimes an abundance of air which can limit the accumulation of organic matter (by accelerating decomposition) and further perpetuate the low level of soil colloids (see chapter 5).

High levels of clay with low levels of OM would have an opposite effect (i.e. a deficiency of air), causing problems associated with anaerobic conditions. The CEC in such a soil might be very high, but the lack of atmosphere in the soil would limit the amount and type of organisms living and/or growing in the area, causing dramatic changes to that immediate environment. Oxidized compounds such as nitrates (NO_3) and sulfates (SO_4) may be reduced (i.e. oxygen is removed) by bacteria that need the oxygen to live. Also the nitrogen and sulfur could be lost as available plant nutrients. Accumulation of organic matter is actually increased in these conditions because the lack of air slows down decomposition. Eventually, enough organic matter may accumulate to remedy the situation, but it could take decades or even centuries.

The CEC of a soil is a value given on a soil analysis report to indicate its capacity to hold cation nutrients (see figure 7-2). However, CEC is not something that is easily adjusted. It is a value that indicates a condition or possibly a restriction that must be considered when working with that particular soil. Unfortunately, CEC is not a packaged product. The two main colloidal particles in the soil are clay and humus and neither are practical to apply in large quantities. Compost, which is an excellent soil amendment, is not necessarily stable humus. Over time, compost may become humus but the end product might only amount to 1-10% (in some cases, less) of the initial application.

Table 7-2 gives an idea of how thick each cubic yard of compost will spread on a 1,000 square foot area. Remember, each percent of organic matter in the soil is equal to over 450 pounds per 1,000 square feet (20,000#/acre). Compost normally contains about forty to fifty percent OM on a dry basis, and weighs approximately 1,000 pounds per cubic yard (depending on how much moisture it contains). If the moisture level is fifty percent, it would take two cubic yards of compost per thousand square feet to raise the soil OM level one percent (temporarily). Obviously, this is not something that can or should be done overnight.

Table 7-2

CUBIC YARDS	DEPTH
.38	1/8"
.75	1/4"
1.5	1/2"
3.1	1"
6.2	2"
9.3	3"

Natural/Organic nitrogen sources, in general, will do more to raise or preserve the level of OM than synthetic chemicals because of the biological activity that they stimulate. Colloidal phosphate contains a natural clay and is often used to condition sandy soils with a low CEC.

If a soil has a very low CEC, adjustments can and should be made but not solely because of the CEC. A soil with a very low CEC has little or no clay or humus content. Its description may be closer to sand and/or gravel than to soil. It cannot hold very much water or cation nutrients and plants cannot grow well. The reason for the necessary adjustment is not for the need of a higher CEC but because the soil needs conditioning. A result of this treatment is a higher CEC.

MILLIEQUIVALENTS (meqs)

CEC is measured in Milliequivalents (meqs) per one hundred grams of soil. An equivalent is actually a chemical comparison. It is a measurement of how many grams of a substance (element or compound) it takes to either combine with or to displace one gram of hydrogen (H). A meq is simply one thousandth of an equivalent.

It sounds complex but it really isn't. Picture a train with one empty seat. The capacity of that seat is one person. If Joe, who weighs 240 pounds, is polite enough to give up the seat to Ann, who weighs 120 pounds, the capacity of the seat hasn't changed, just the weight of the passenger occupying it. When measuring CEC, each ion is a passenger on the soil colloid. Unfortunately, unlike commuters, ions are far too small to count; therefore, their numbers must be calculated by using known factors such as atomic weight and electro-magnetic charge.

Hydrogen has an atomic weight of one and has a valence of $^+1$ (i.e. a positive charge of 1). Calcium (Ca), for example, has an atomic weight of forty and a valence of $^+2$. H and Ca are both cations, and

Table 7-3

ELEMENT or COMPOUND	ATOMIC WEIGHT	VALENCE	EQUIVALENCE
Hydrogen	1	1	1
Magnesium	24	2	12
Calcium	40	2	20
Potassium	39	1	39
Ammonium	18	1	18

they would not combine, so the equivalence here is in terms of H displacement. Since Ca has twice the charge as H (i.e. it occupies two seats on the colloidal express), only half of its atomic weight is needed to displace the atomic weight of H. Therefore, it would take twenty grams of Ca to displace one gram of H. A CEC with a value of 1 meq/100 gr means that each one hundred grams of soil can magnetically hold either one milligram (mg) of H or twenty mg of Ca (more likely some combination of both). Table 7-3 shows equivalent values of common soil cations.

The CEC is an overall measurement and is calculated from the amount of exchangeable base cations (i.e. Ca, Mg and K) and meqs of H found during soil analysis. In a test where exchangeable cations are measured in parts per million (ppm), the equivalence of the cation is multiplied by ten (because the ratio between parts per million is ten times greater than the ratio of milliequivalents to 100 grams) and then divided into the ppm found in the analysis. For example, if 125 ppm potassium were found in the soil test results, its equivalence in meq/100gr would be:

125 divided by 390 = 0.3 (rounded)

125 is the ppm found in the test and 390 is the equivalent of potassium (39) multiplied by 10.

Once the ppm of Ca, Mg and K are calculated into meq/100gr, they are added together. This total constitutes the portion of the soil's CEC currently occupied by the base cations. Meqs of H, are calculated from the buffer pH test and is then added to the total from the bases to arrived at the soil's overall CEC.

EXAMPLE IN TEST # 1 (from figure 7-2):

BASE	PPM FOUND	Meq/100gr
Potasium	139	.4
Magnesium	187	1.6
Calcium	640	3.2

Total Bases	5.2	
Hydrogen	.5	*not shown*
CEC	5.7	

EFFECTIVE CEC

Labs that offer a CEC value have calculated all of this in advance. However, if a lab does not run a Buffer pH test, it cannot include H in its CEC calculations. The University of Vermont (UVM), for example, offers what they call *Effective CEC* (ECEC) which is only calculated from the exchangeable base cations found. Since UVM calculates its lime recommendations from an analysis of exchangeable aluminum, it has no H^+ value to plug into this formula. If calculating the CEC is necessary, its lime recommendation will act as a H^+ value. For example, if UVM recommends three tons (6,000 lbs) of lime per acre, each 1,000 lbs of lime will neutralize 1 meq of H. Therefore, three tons will neutralized 6 meq of H. According to UVM (and others), the CEC of a soil is pH dependent and should fluctuate with variances in pH. It is UVM's opinion that ECEC is a more accurate measure of cation exchange capacity.

ANIONS

Anions are atoms or groups of atoms that carry a negative electromagnetic charge. Because they have the same polarity as soil colloids, they cannot be held or exchanged by clay or humus particles. The three anionic compounds most related to plant nutrition are nitrate (NO_3), phosphate (PO_4) and sulfate (SO_4). Phosphate ions bond easily with many different soil elements or compounds, and do not migrate very far before finding a friend to live with. However, topical applications of available phosphate can be eroded with normal surface runoff. This runoff causes nutrient loss and possible eutrophication (biological pollution) of waterways.

Nitrates and sulfates are susceptible to leaching. Aside from the issue of ground water pollution, the loss of these nutrients when applied as inorganic compounds, can be significant. Dissolved nitrates and sulfates in soil solution can be held more efficiently by soils with a higher CEC simply by virtue of the increased water holding capacity. However, in heavy clay soils where the oxygen supply is limited, reduction (i.e. the biological or chemical removal of oxygen from compounds) can occur, because bacteria transforms the nutrients into gases that can escape into the atmosphere.

Cations, anions and the exchange system in soils is a crucial component in the cycles and chains of life on earth. Its importance is comparable to a vital organ which, by itself, does not sustain life, but without it, life could not exist.

PHOSPHORUS

Most labs report both available and reserve phosphorus and usually utilize symbols such as VL (very low), L (low), M (medium), H (high) and VH (very high) to indicate how the results of the test compare to what they consider acceptable levels. These symbols are used throughout the soil analysis report.

Phosphorus (P) occurs in the soil as a phosphate (PO_4) ion. Phosphate ions have a negative charge (anionic) and do not cling to clay or humus particles. Available phosphate can be lost through surface run off but otherwise does not migrate in the soil because phosphate ions can easily combine with other elements in the soil or it can be utilized by soil organisms and plants. Plants utilize phosphorus as a phosphate combined with hydrogen (phosphoric acid).

Phosphorus tests are inherently inaccurate because there are so many variables in the soil that affect the availability of phosphate. However, if the lab finds deficiencies of P then an application of some type of phosphate is usually warranted. The problem occurs when the lab finds acceptable levels, but soil conditions are such that it cannot become available. The activity of microorganisms is instrumental in freeing up phosphate by 1) mineralizing organic phosphate from dead organisms and, 2) by chemical reactions with soil compounds that contain phosphate. Good levels of soil organic matter are the key to a healthy population of microbes and adequate levels of available phosphate.

SOIL LEVELS OF EXCHANGEABLE CATIONS

Ca, Mg and K levels are usually tested for by standard procedures, but many labs will report them differently. For example, K (potassium) may be reported as K_2O (potash) instead, or all three cation elements may be reported in pounds per acre (ppa) instead of parts per million (ppm). This will significantly change the way application rates of nutrients are calculated. Make sure the lab indicates what measurement units it is using.

BASE SATURATION

Base saturation is simply the balance or ratio of base cations in the soil. Base saturation should be used as a guideline. Labs that offer base saturation percentages will also give normal ranges that those values should fall within. A & L suggests the following ranges:

K = 2 - 7%   Mg = 10 - 15%   Ca = 65 - 75%

Values significantly above the accepted range should not cause concern unless they are creating an imbalance in the form of deficiencies or excessively high pH. Only the deficiencies should be addressed. Knowing the CEC and base saturation tells two very basic facts about the soil. First, it tells how much Potash, Magnesium and Calcium the soil can hold and, second, if the proper balance of those nutrients exist. Table 7-5 shows where cation nutrient levels should be to achieve balanced base saturation at different CEC's. This table is taken from the A & L Agronomy Handbook for Soil and Plant Analysis which is available from A & L Laboratories for only $5.00. The information contained in the 133 pages of this manual is easily worth ten times that amount.

TEXTURE ANALYSIS

Texture analysis is done by determining the percentage of sand, silt and clay in a given soil. Those findings are plugged into a texture analysis triangle (figure 7-9) to determine soil classification. The triangle is used by extending lines from the appropriate starting points, parallel to the side of the triangle, counterclockwise to the side where the line began. For example, if a soil texture analysis discovered 40% sand, 40% silt and 20% clay, the first line would begin on the 40 mark of the "percent sand" side of the triangle, drawn parallel to the "percent silt" side. The second line begins at the 40 mark on the "percent silt" and is drawn parallel to the "percent clay" side. Where those two lines meet is the texture classification of the soil. The third line may be drawn to complete the triangulation, but it is not necessary.

A simple texture analysis can be conducted by collecting a soil sample in a glass jar or test tube and mixing it with an equal volume of water. The dry sample should be measured (linear from the bottom of the jar to the top of the sample or volumetrically) before water is

TEXTURE ANALYSIS CHART

Figure 7-9

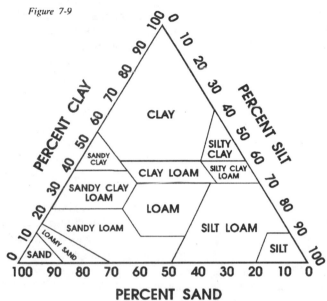

PERCENT SAND

added. After adding water, shake the container vigorously and then place in a location that will allow it to be undisturbed for twenty four hours. The sand particles will settle to the bottom and the silt above that. Clay will settle on top but can take up to two weeks to precipitate out of suspension, which is why it is important to measure the dry sample before adding the water. If the volume of the original sample is known, it is easy to determine the percentage of sand and silt by measuring each layer and calculating what portion it is of the whole sample. Once the percent sand and silt is determined, use the texture analysis chart (see Figure 7-9) to classify the sample.

INTERPRETING THE ANALYSIS

Given the large number of different tests available from an even larger number of laboratories, it would be impossible to outline a blanket method of interpretation for every one of them. However, if we interpret the analysis for sample #1 (top line) shown in figure 7-2, we can get a rough idea of how to apply this method to the test your lab does.

Table 7-4

BUFFER pH	MINERAL SOILS	ORGANIC SOILS
7.0	0	0
6.9	0	0
6.8	1	0
6.7	1.5	0
6.6	2.0	0
6.5	2.5	0
6.4	3.0	1.0
6.3	3.5	2.0
6.2	4.0	2.5
6.1	4.5	3.0
6.0	5.5	4.0
5.9	6.0	4.5
5.8	6.5	5.0
5.7	7.0	5.5
5.6	8.0	6.0
5.5	9.0	6.5

1. The soil pH indicates the need for lime and the buffer pH indicates how much lime to apply (see table 7-4). This table appears on the back of every A&L Soil Analysis Report. Mg levels are very high so dolomite (hi-mag lime) should not be used if other liming materials such as calcite or aragonite are available. Lab recommendations for lime are often higher than necessary. Cautious applications of 50-75% of the labs recommendation in conjunction with re-testing after six months may be well advised. Applications without incorporation (tilling, harrowing or plowing) should not exceed 2.5 tons of lime per acre (115 lbs. per 1,000 sq. ft.). Too much lime at a time can upset the ecology of a soil.

2. The soil organic matter level is adequate.

3. The soil CEC is not out of proportion to the level of OM found, indicating only a small amount of clay is in this soil.

4. The soil phosphorus levels are very high.

5. The soil potassium level is very high. If potassium were needed, Table 7-5 (from the A&L Handbook) shows how to calculate an application rate. With a CEC of 6.2, the amount of K in ppm should be over 117. Since 139 ppm already exists, more is not necessary. If less than 117 ppm were found, the difference would be applied after being converted to pounds per acre. Noted on the bottom, right side of the analysis report (not shown in figure 7-2), is the conversion of potassium in ppm to potash in ppa, (multiply by 2.4).

NOTE: Some application rates may be too much nutrient to put down all at once. Split applications (apply half at a time) might be

| Table 7-5 | | From: A&L Agronomy Handbook | |
CEC	POTASSIUM 2-5%	MAGNESIUM 10-15%	CALCIUM 65-75%
30	292	360	3900
29	284	348	3770
28	274	336	3640
27	264	324	3510
26	254	312	3380
25	244	300	3250
24	234	288	3120
23	224	275	2990
22	215	263	2860
21	205	252	2730
20	195	240	2600
19	192	236	247
18	187	230	2340
17	182	225	2210
16	176	218	2080
15	170	210	1950
14	164	202	1820
13	158	193	1690
12	152	183	1560
11	147	172	1430
10	141	160	1300
9	135	148	1170
8	129	135	1040
7	123	121	910
6	117	106	708
5	108	90	650
4	85	75	520

more appropriate. There is no general rule that applies to all nutrients as far as maximum application rates are concerned. Common sense is usually an accurate barometer but, when in doubt, contact the lab for guidance.

6. The base saturation shows less than adequate levels of Ca confirming the need for lime application.

Correcting deficiencies in a soil with a very high CEC is more difficult because the capacity is so large. Imagine yourself working a hand pump with two buckets to fill. One holds a gallon and the other is a fifty five gallon drum. Not only does the drum take much more material, but it also takes a lot more energy to fill it. However, if both containers are filled to capacity and the contents of both are used at the same rate, the advantages of the larger container are obvious.

NITROGEN

Nitrogen (N), which is considered to be the most used soil nutrient, has not been mentioned until now, and it did not appear on our list of important information offered by labs. Nitrogen is a little like New England weather. It changes so rapidly that it is impossible to get an accurate reading that means anything. Most labs offer tests for nitrate nitrogen (NO_3) but, in order to get an accurate result, they suggest that the sample is dried or frozen immediately to stop the biological activity in the sample. Estimated Nitrogen Release (ENR) can be calculated from the level of organic matter found in the soil. However, it is difficult to predict how much of that nitrogen will become available because of the large number of variables that affect its release. On the other hand, nitrogen applications should not be based strictly on plant needs. For example the N that turf receives from organic matter, especially when clippings are recycled, is significant. Infusions of soluble nitrate or ammoniated fertilizers in excessive doses can do considerably more harm than good in the soil and can cause pollution in ground water, not to mention the money wasted on these materials.

MICRO-NUTRIENTS

Although most plants do not have an especially high need for micro-nutrients, all plants need some. Any deficiency of a nutrient, no matter how small an amount is needed, will hold back plant development. Figure 7-10 illustrates the equal importance of all essential nutrients.

Testing for micro-nutrients can get expensive but if problems exist that are not explained by a standard test, it may be necessary to spring for it. Many labs hold onto the original sample for thirty days after they have analyzed it so drawing new samples may not be necessary. A simple phone call should get the test performed. Most micro-nutrients are cations, but labs do not report results in relation to CEC. The levels of micro-nutrients needed in the soil are so small that virtually any CEC will hold what is necessary.

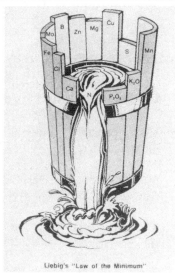

Liebig's "Law of the Minimum"

White 1982

Great care should be taken when correcting micro-nutrient deficiencies. There is a very fine line between too little and too much. The lab that does the testing can make recommendations but their recommendations are only as good as the sample taken. Micro-nutrient tests are easily adulterated by rusty or corroded tools. Because of the high zinc content, galvanized tools or containers should never be used regardless of what their condition may be.

SUMMARY

Recommendations from most labs for pounds of nutrient per acre or per thousand square feet only apply to plant needs and can prove to be useless or even injurious to the other organisms in the soil. However, if nutrients are added with consideration of the entire soil system, then the whole solar powered, biological crop growing machine can benefit.

It is usually not necessary to fine tune the soil. It does not respond like a high performance engine where subtle adjustments can tweak out another two to three more horsepower. Remember, the results of a soil test are an average of the area being evaluated. If inputs contain the raw materials needed for natural soil system mechanisms, then the system will function correctly.

IMPORTANT POINTS TO REMEMBER:

Good samples - a lab analysis is useless if poor sampling procedures are implemented.

Choose a lab that gives the type of information needed on a timely basis.

Consider the needs of the soil as well as those of the plants.

A LIST OF SOME SOIL TESTING LABORATORIES

A & L Agri. Labs. of Memphis 411 North 3rd St.
Memphis, TN 38105-2723 901/527-2780

A & L Eastern Agricultural Labs 7621 Whitepine Rd.
Richmond, VA 23237-2296 804/743-9401

A & L Great Lakes Labs, Inc. 3505 Conestoga Drive
Fort Wayne, IN 46808-4414 219/483-4759

A & L Mid West Labs., Inc. 13611 "B" St.
Omaha, NE 68144-3693 402/334-7770

A & L Plains Agric. Labs, Inc. PO Box 1590, 302 34th St.
Lubbock, TX 79408-1590 806/763-4278

A & L Southern Agri.Labs., Inc. 1301 W. Copans Rd.,
Bldg. D. #8
Pompano Beach, FL 33064 305/972-3255

A & L West. Agricultural Labs 1010 Carver Rd.
Modesto, CA 95350-4732 209/529-4080

Agri. Labs Inc. 204 East Plymouth
Bremen, IN 46506

Agrico-Chemical Co. PO Drawer 639
Washington Court House, OH 43160

Agricultural Testing Lab. Hills Building, UVM
Burlington, VT 05405 802/656-3030

Analytical Services Laboratory 8 Nesmith Hall, UNH
Durham, NH 03824

Brookside Research Labs. 308 South Main St.
New Knoxville, OH 45871

Cal Mar Soil Testing 130 South State St.
Westerville, OH 43081

Cooperative Ext. Publications University of Illinois
Urbana, IL 61801

Cooperative Extension Service 2120 University Ave.
Berkeley, CA 94720

Farm Clinic 923 Robinson St., PO Box 3011
West Lafayette, IN 47906

Freedom Soil Lab PO Box 1144G
Freedom, CA 95019 408/724-4427

Geophyta 2685 County Rd 254
Vickery, OH 43464

Harris Laboratories, Inc. PO Box 80837, 624 Peach St.
Lincoln, NE 68501

Indiana Frm Bureau Central Lab. 2435 Kentucky Ave.
Indianapolis, IN 46204

Iowa Testing Laboratory Highway 17 North, Box 188
Eagle Grove, IA 50533

LaRamie Soils Service PO Box 255
Laramie, WY 82070

Merkle Laboratory Pennsylvania State University
University Park, PA 16802

Missouri Western State College MEY S.T.L.,ET 114,4525
Downs Dr St. Joseph, MO 64507

Na-Churs 421 Leader Street
Marion, OH 43302

Perry Laboratory 471 Airport Blvd.

Watsonville, CA 95076 408/722-7606

Pike Lab Supplies RR #2, Box 92
Strong, ME 04983 207/684-5131

Plant and Soil Analysis Lab. Agronomy Dept., Purdue Univ.
W. Lafayette, IN 47907

Research-Ext. Analytical Lab. Ohio Agricultural R & D Center
Wooster, OH 44691

Rutgers' Soil Testing Laboratory Cook College, Rutgers Univ.
New Brunswick, NJ 08903

Soil & Plant Laboratory, Inc. PO Box 153
Santa Clara, CA 95052-0153 408/727-0330

Soil Test. Lab./Dept. of Agron. Bradfield & Emerson Halls
Ithaca, NY 14853

Soil Testing Laboratory University of Connecticut
Storrs, CT 06268

Soil Testing Laboratory University of Maine
Orono, ME 04473

Soil Testing Laboratory Clemson University,
Clemson, SC 29631

Soil Testing Laboratory University of Georgia
Athens, GA 30602

Soil Testing Laboratory University of Massachusetts
Amherst, MA 01002

Soil Testing Laboratory 210B Woodward Hall, URI
Kingston, RI 02881

Soil Testing Laboratory Michigan State University
East Lansing, MI 48823

Soil Testing Laboratory North Carolina State University
Raleigh, NC 27695

Soil Testing Laboratory Texas A & M University
College Station, TX 77843

Terra Analytical Services 2622 Baty Road
Elida, OH 45807

WDHIC Soil & Forage Center 106 North Cecil Street
Bonduel, WI 54107

Sources:

Albrecht, W.A. 1938, Loss of Organic matter and its restoration. U.S. Dept. of Agriculture Yearbook 1938, pp347-376

Arshad, M.A. and Coen, G.M. 1992, Characterization of soil quality: Physical and chemical criteria. American Journal of Alternative Agriculture v7 #1 and 2, 1992 pp 25-31. Institute for Alternative Agriculture, Greenbelt, MD

Brady, N.C. 1974, The Nature and Properties of soils. MacMillan Publishing Co. Inc. New York, NY

Chu, P. 1993, Personal communication. A&L Eastern Agricultural Laboratories. Richmond, VA

Ehrlich, P.R. and Ehrlich, A.H. 1990, The Population Explosion. Simon and Schuster. New York, NY

Gershuny, G. and Smillie, J. 1986, The Soul of Soil: 2nd Edition. Gaia Services. St. Johnsbury, VT

Jenny, H. 1941, Factors of Soil Formation. McGraw - Hill Book Co. New York, NY

NRAES, 1992, On Farm Composting Handbook. Northeast Regional Engineering Service #54, Cooperative Extension. Ithaca, NY

Parnes, R. Fertile Soil: A growers Guide to Organic & Inorganic Fertilizers. Ag Access, Davis, CA

Seyer, E. 1992, Sustaining a Vermont Way of Life: Research and education in Sustainable Agriculture. University of Vermont. Burlington, VT

White, W.C. and Collins, D.N. (Editors) 1982, The Fertilizer Handbook. The Fertilizer Institute. Washington, DC

Chapter 8

RELATIONSHIPS

Love is an emotion; a feeling of endearment. It is a feeling which is slightly different for everyone. Like love, all of the other feelings we have in our relationships with others are hard to explain. Sometimes they are good and sometimes they are not, but we must accept them, without completely understanding them. As we grow, we learn to adapt our feelings to the many different relationships that we encounter, just as the soil relates to the different *edaphic factors* it encounters.

All of our feelings are derived from instincts that have evolved from a basic need to survive. These instincts are like patterns that emanate throughout the biological and physical components of the planet. The hundreds of similarities between human relationships and those in the soil are not coincidences.

Just like cations and anions, we have stronger attractions to some relationships than we do for others. Sometimes our polarity for one another is so opposite (or alike), that we instinctively know the chance of a good relationship is remote at best.

This same polarity governs us like molecules of water. We group together in cohesive societies and enforce our own rules to regulate our flow.

The most basic and driving forces in our lives are for sustenance and to reproduce. Those same instincts control all the biological functions in the soil.

Our minds collect the wisdom of generations and we use it to produce, just as the slow evolution and accumulation of humus helps produce the living wealth of the planet.

When we purchase essential supplies such as food and clothing, our relationship with the merchant is symbiotic, like the relationship that mycorrhizae fungi have with trees and other plants. We get sustenance and give it at the same time. And, just as plants compete for water, nutrients, and light, so do we for jobs, spouses and more comfortable habitats. When we are stressed either physically and/or mentally we become susceptible to problems, like the plants and other organisms in the soil. We sleep to rest and rejuvenate our body and spirit just as the field left fallow rebuilds its energy reserves. We work at what we do to survive; like plants, animals, microbes and all other living things do. And, although our definition of survival is different than what we apply to other species, we all struggle, nonetheless.

There are many more similarities between our relationships and those that exist in the soil. However, we are unique. The human being is the most cunning and imaginative of all species. We can reason beyond instinct into technologies that dwarf our ancestor's wildest dreams. We have harnessed energy of all kinds and have enslaved it to provide us with comfort (or to take it from others). We change our clothes, our hair, our schedule and even our life-styles at a whim, while the soil plods on as it has for millions of years. We choose to dominate rather that tolerate annoyances from each other and our environment. The delicate balance that exists on earth is being slam dunked by our species and is quickly being drained of resources that were designed to last an eternity.

Many of us neglect ourselves as much as or more than we neglect the soil. We pay little attention to what or how much we eat and forego exercise as a meaningless annoyance. At some point we discover that repairing the damage from our neglect is much more difficult than the maintenance would have been. The same is true with our soil. The only difference is that when the soil is worn out, we can just move on. However, the space on earth is finite, and soon there will be no place to move on to.

Our ability to care and respect is the key to all of our relationships. If one has no respect for himself, it is unlikely he will care for anyone or anything else. If, on the other hand, we have great self respect, then we have the potential to do great things.

This care and respect is a crucial component of our existence as *edaphic factors*. Yes, we are factors of our own environment. We affect and are effected by all that goes on around us. We must understand that every action we take will eventually have an effect on the entire ecosystem. We cannot know if everything we do is right, but we must begin to care. If we as a species ignore our link to the soil and relate to it without any care and respect, then we unwittingly blaze that fateful trail toward extinction.

GLOSSARY

This glossary is intended to not only provide definitions of terms used within the text of **Edaphos**, but also to provide some definitions of the terms used in many of the sources listed at the end of each chapter.

Abiotic - Not living.

Acid - Any substance that can release hydrogen ions in a solution.

Actinomycetes - Decay microorganisms that have a fungus like appearance but, like bacteria, do not contain a well defined nucleus.

Adsorb - See Adsorption

Adsorption - The adherence of one material to the surface of another via electro-magnetic forces, e.g. dust to a television screen.

Adventitious rooting - roots emanating from above ground plant parts.

Aerobic - Needing oxygen to live.

Alkaline - Refers to substances with a pH greater than 7.

Allelopathic - Usually refers to the negative influence a plant has on other plants or microorganisms.

Amorphous - Without consistency in its structure or form.

Anaerobic - Refers to an environment with little or no oxygen or organisms that require little or no oxygen to live.

Antagonist - Any organism that works against the action of another.

Anthropogenic - Caused by human action.

Apatite - A natural phosphate mineral.

Aragonite - Calcium carbonate (lime) formed by shellfish.

Arthropod - Refers to a group of animals with segmented bodies and exoskeletal structure such as insects, spiders and crustaceans.

Assimilation - Digestion and diffusion of nutrients by an organism for growth and/or sustenance.

ATM - Atmosphere, a measurement of pressure of a gas.

Atmosphere - Refers to the naturally existing gases of any given environment. Also a measurement of pressure (see ATM).

Autoclave - A machine that, with heat and pressure can react gases with other materials. It is also used to sterilize tools and equipment.

Autotrophs - organisms that can synthesize organic carbon compounds from atmospheric carbon dioxide utilizing energy from light or chemical reactions.

Bases - See Base cation.

Base cation - A positively charged ion, historically belonging to the earth metal family e.g. potassium, magnesium, calcium, etc.

Biomass - The accumulative mass of all living things in a given environment.

Biotic - Pertaining to life or living organisms.

Botany - The study of plants.

Calcite - Calcium carbonate (lime)

Carbon : Nitrogen ratio - a ratio measured by weight of the number of parts carbon to each part nitrogen e.g. 10:1, 50:1, etc.

Carbohydrates - A group of organic compounds that include sugars, starch and cellulose.

Carnivores - Organisms that consume animals or insects for sustenance.

Cation - An ion of an element or compound with a positive electromagnetic charge.

Cation exchange capacity - The total amount of exchangeable cations that a given soil can adsorb.

Cellulose - The most abundant organic compound on earth found mostly in the cell walls of plants.

Chelation - The combination of metal (inorganic) and organic ions into a stable compound sometimes referred to as a chelate.

Colloids - Very small soil particles with a negative electo-magnetic charge capable of attracting, holding and exchanging cations.

Cultivar - Refers to a specific plant variety.

Detritus - The detached fragments of any structure, whether biotic or abiotic that are decomposing or weathering.

Dolomite - Calcium, magnesium carbonate (magnesium lime).

Edaphic - Refers to factors, such as soil structure, climate, fertility and biological diversity that influences the growth of plants.

Edaphos - Greek word meaning soil.

Entomo - Prefix pertaining to insects.

Enzymes - A group of proteins that hasten bio-chemical reactions in both living and dead organisms.

Eutrophy - Refers to the excessive nutrient enrichment of ponds or lakes causing the accelerated growth of plants and microorganism and the depletion of oxygen.

Exude - The release of substances from cells or organs of an organisms.

Faunal - Pertaining to microscopic or visible animals.

Fecundity - Refers to the reproductive capabilities of an organism.

Floral - Pertaining to plants or bacteria, fungi, actinomycetes, etc.

Free oxygen - Gaseous oxygen not bound to other elements

Furrow slice - Plow depth of approximately 6 - 7 inches.

Geoponic - Pertaining to agriculture or the growing of plants on land.

Gustation - or gustatory refers to an organisms sense of taste.

Hemicellulose - A carbohydrate resembling cellulose but more soluble; found in the cell walls of plants.

Herbivores - Organisms that consume plants for sustenance.

Heterotrophs - Organisms that derive nutrients for growth and sustenance from organic carbon compounds but incapable of synthesizing carbon compounds from atmospheric carbon dioxide.

Humification - The biological process of converting organic matter into humic substances.

Humology - The study of humus.

Hydrolysis - The reaction of hydrogen (H) or hydroxyl (OH) ions from water with other molecules usually resulting in simpler molecules that are more easily assimilated by organisms.

Hyphae - A microscopic tube that is a basic component of most fungi in their growth phase.

Ion - Any atom or molecule with either a positive or negative electromagnetic charge.

Kairomone - A chemical substance produced by an organism (e.g. plants or insects) that attracts another species or gender to it. Example: 1) Fruit produces kairomones that attract certain insects. 2) Many insects produce kairomones called pheromones that attract the opposite sex.

Ligand - An anionic compound. A molecule of a compound that carries a net negative electro-magnetic charge that can bond with cations.

Lignin - A biologically resistant fibrous compound deposited in the cell walls of cellulose whose purpose is for strength and support of stems, branches, roots, etc.

Macro - A prefix meaning large.

Meso - A prefix meaning middle.

Metabolism - The biological and chemical changes that occur in living organisms or the changes that occur to organic compounds during assimilation by another organism.

Metabolite - A product of metabolism or a substance involved in metabolism.

Meteorology - The study or science of the earths atmosphere.

Methodology - A system or the study of methods.

Micelle - (Micro-cell) A negatively charged (colloidal) soil particle most commonly found in either a mineral form (i.e. clay) or organic form (i.e. humus).

Micro - A prefix meaning small, usually microscopic.

Mineralized - The biological process of transforming organic compounds into non-organic compounds (minerals) e.g. mineralization of protein into ammonium.

Mineralogy - The study or science of minerals.

Mitigate - To lessen.

Molecule - The smallest particle of a compound that can exist independently without changing its original chemical properties, e.g. one molecule of water contains one atom of oxygen and two of hydrogen (H_2O).

Monoculture - The cultural practice of growing only one variety of crop in a specific area every season without variance.

Morphology - The study or science of the form or structure of living organisms.

Mucilage - compounds synthesized by plants and microbes that swell in water, taking on a gelatinous consistency, that function to maintain a moist environment.

Myco - A prefix that refers to fungi.

Nitrification - A process performed by soil bacteria that transforms ammonium nitrogen into nitrite and, finally, nitrate nitrogen. Nitrate is the form of nitrogen most often used by plants.

OM - Abbreviation for organic matter.

Oxidation - Usually refers to the addition or combination of oxygen to other elements or compounds.

Oxidize - To add oxygen. See oxidation.

Parent Material - The original rock from which a soil is derived.

Pathogen - any microorganism that causes disease.

Pedology - The study or science of soils.

Pedosphere - The top layer of the earths crust where soils exist.

Phenology - The study or science of biological phenomena and its relationship to environmental factors.

Pheromone - A chemical produced by an insect or other animal that attracts another member of the same species, usually of the opposite sex.

Physiology - The study or science of the biological functions and/or activities of living organisms.

Phyto - A prefix referring to plants.

Phytotoxic - A substance that is toxic to plants.

Porosity - Refers to the spaces between soil particles.

Rhizosphere - The area of soil in immediate proximity to roots or root hairs of plants.

Saprophyte - an organism that can absorb nutrient from dead organic matter.

Senescense - The aging process.

Situ - Refers to situations. Example: organisms in situ may respond differently to a stimulus than they would in a laboratory.

Sodic soil - Any soil with sufficient sodium to interfere with the growth of plants.

SOM - Abbreviation for soil organic matter.

Steward - A person who manages or cares for property of another. In agriculture, the term can refer to someone who cares for his own land but believes that ownership does not entitle one to dispose of the soil's resources for his/her own personal gain.

Substrate - Material used by microorganisms for food.

Symbiotic - A relationship between two organisms, usually obligatory and often of mutual benefit.

Synergy - Where the activities or reactions of two or more organisms or substances are greater than the sum of the agents acting separately.

Taxonomy - The science of classification.

Tectonic - Pertaining to the structure and form of the earths crust.

Texture analysis - An analysis of soil particles determining the percentages of sand, silt and clay.

Throughfall - Moisture or precipitation that drips from above ground plants, such as trees, to the ground. Throughfall is thought to contain some substances leached from leaf surfaces.

Topography - Pertaining to the specific surface characteristics of a given landscape.

Trophic levels - Levels of consumers within a food chain in relation to producers of organic nutrient such as plants e.g. producers - primary consumers - secondary consumers - tertiary consumers - decay organisms.

Valence - A measurement of how many electrons an atom or molecule can share in a chemical combination. A positive valence indicates electrons offered in a chemical bond whereas a negative valence is the number of electrons that can be accepted.

Volatile - Refers to substances that can easily change, oftentimes into a gas.

Index

A

Acid rain   12, 27, 79
Acidity   143
Acids   9, 11, 87
Actinomycetes   34
Adaptation   75
Adhesion   54
Aeration   45, 90, 91, 93
Aerobic   90
Aflatoxins   117, 119
Air   90, 100
Air pollution   79
Algae   110
Alkalinity   143
Alkaloids   74
Allelopathic compounds   77
Allophanic soils   44
Aluminum   15, 118, 148, 156
Amino acids   71, 110, 121
Ammoniated fertilizers   162
Anaerobic   92
Anaerobic bacteria   90
Animal
  By-products   119
  Feeds   119
  Manures   115
  Pests   120, 121
  Residues   108, 116
  Tankage   119
Anions   156
Antagonism. *See* Antagonists
Antagonists   70, 78
Antibiotics   73
Apatite   121, 122, 125

Aragonite   131, 160
Aral Sea   26
Ash   99, 125, 140
Assets   101
Atacama Desert   117
ATM   57
Atmosphere   7, 18, 19, 53, 59, 110, 111, 112, 117, 121,
      129, 145, 156
Atmospheric
   Carbon   67
   Nitrogen   112
   Pollution   28
Atomic weight   154
Atoms   147
Autoclaving   118
Autotrophic organisms   18, 39, 41, 110, 145
Available
   Phosphate   133, 157
   Phosphoric acid   133
   Phosphorus   143

B

Bacteria   34, 46, 73, 85, 87, 90, 93, 97, 98, 110, 112,
      113, 118, 125, 127, 131
Beneficial Bacteria. *See* Bacteria
Baking soda   70
Bare soil   96
Basalt   128
Base cations   43, 150
Base saturation   143, 158
Beneficial organisms   79, 146
Big eyed bugs   68
Bio-stimulant   130
Biological equilibrium   76, 86
Black Rock phosphate   122
Blood meal   116
Bone meal   121
Boron   21
Buffer pH   142, 143

Bulk soil density   87

C

C:N ratio   96, 97, 100, 110, 113, 114, 115
Calcined Rock Phosphate   123
Calcite   19, 160
Calcium   15, 21, 42, 43, 85, 86, 122, 123, 124, 126, 131,
    134, 154, 158
  Carbonate   131, 134, 151
  Magnesium carbonate   131
  Phosphate   124
  Sulfate   124, 132
Capillaries   55
capillary action   55
Carbohydrates   111
Carbon   21, 41, 42, 67, 86, 96, 98, 110, 111, 112, 113,
    115, 117, 119, 121, 123, 125, 147
  Compounds   85, 108, 110
  Cycle   88, 111, 117
  Dioxide   10, 12, 18, 19, 21, 25, 36, 39, 45, 47, 88,
    99, 111, 117, 131, 145
Carbon:Nitrogen (C:N) Ratio. *See* C:N ratio
Carbonates   131, 132, 151
Carbonation   12
Carbonic acid   10
Cation exchange   41, 150
Cation exchange capacity (CEC)   143, 149, 152, 156
Cations   15, 156
Cellulose   37, 38
Certified organic   107
CFC's (Chloroflorocarbons)   27
Chelated trace elements   128, 130
Chelates. *See* Chelated trace elements
Chemical
  Defense compounds   73
  Nitrogen   110
  Weathering   9
Chilean nitrate   108, 117, 120
Chinch bugs   68, 74, 132

Chlorine   21, 134
Chlorophyll   21
Chromium   118
Clay   14, 44, 85, 86, 92, 148
Clay soils   156. *See also* Clay
Climate   72, 75
Coal   27
Cobalt   21
Cocoa meal   119, 121
Coffee wastes   119
Cohesion   54
Cold   79
Colloidal   15
   Clay   123
   Complexes   149
   Exchange sites   150
   Phosphate   92, 123, 154
Colloids   41, 150
Compaction   77, 85, 100
Competition   69, 77, 78
Competitors   70. *See also* Competition
Compost   72, 89, 90, 92, 93, 101, 114, 121, 125, 129
Composted manures   116
Composting   89
   Vessels   94
Compound   147
Condensation   60
Copper   21, 148
Core aeration   91
Cottonseed meal   119
Cover crops   78, 129
Crop rotations   78
Cultivation   78, 79
Cultivation equipment   91
Cultural practices   89, 101
Cycle of nitrogen   113. *See also* Nitrogen cycle

D

DAP (diammonium phosphate)   122

Decay of organic matter   89. *See also* Humus
Decay organisms   96, 100. *See also* Humus
Decomposition bacteria   114
Defense compounds   73
Defenses   79
Derived from list   134
Desertification   28
Destruction of ecosystems   23
Disease   69, 72, 86, 112, 130
Diversity   75
Dolomite   19, 132
Dried blood   118
Dried whey   116, 119
Drought   28, 79, 100
   Tolerance   112
Dry leaves   93
Dust bowl days   91

E

Earthworms   87
Edaphic
   Factors   19
   Theory   28
Effective CEC   156
Elemental sulfur   70
Elements   147
Endophytes   74
Energy   19, 25, 39, 54, 74, 79, 86, 98, 110, 111, 112,
      117, 118, 121, 131, 152
ENR   146. *See* Estimated Nitrogen Release (ENR)
Enzymes   9, 11, 71, 87, 110, 128
Epsom Salts   132
Equilibrium   68, 75
Equivalent   154
Erosion   17, 20, 23, 61, 85, 145
Estimated Nitrogen Release (ENR)   146, 162
Ethylene   73
Eutrophication   122, 156
Evapo-transpiration   60

Evaporation   55, 60
Evolution   18, 20
Excess nitrogen   110. *See also* Nitrogen
Excessive porosity   77
Exchange system   156
Exudates   87
Eye ball society   26

F

Fats   20, 37
Feather meal   116, 118
Feathers   118. *See* Feather meal
Feces   36
Fertility   71, 79, 101
Fertilizer labels   134
Fertilizers   26, 107, 115, 117
Fiber   20
Field capacity   56
Field corn   78
Fish
  Emulsion   120
  Meal   120
  Waste   120
Fixation   122
Flowering   78
Food chain   39
Forests   75, 76
Fossil
  Fuels   25
  Shells   122
Free energy   54
Frost   8, 145
Fruit trees   69
Fulvic acid   41
Fungal disease   100
Fungi   34, 67, 68, 73, 119
Fungicides   68

G

Germination 79
Glacier 9
  Glacial dust 9
  Glaciation 9
Glauconite 126
Global climates 79
Grade 133
Granite dust 128
Granulation 132
Gravity 8, 55
Green manures 88, 128, 129
Green Revolution 25
Greenhouse warming 24
Greensand 125, 126, 128
Ground cocoa shells 119. *See also* Cocoa meal
Ground water pollution 79, 112, 156
Growth hormones 130
Grubs 74, 132
Gypsum 124, 132

H

Hair 118
Harrows 91
Health problems 79
Heat 79
Hemicellulose 37
Herbicidal properties 77
Herbicide 66, 77, 78, 129
Herbivores 70, 78
Herbivorous
  Insects 72
  Pests 70
Heterotrophic organisms 19, 41. *See also* Heterotrophs
Heterotrophs 18, 110, 111. *See also* Heterotrophic organisms
Hoof and horn meal 116, 118
Hormone 73
Human Factor 22

Humans   18
Humates   41, 130
Humic acid   41
Humification   35
Humins   41
Humus   20, 23, 33, 85, 87, 89, 98, 130, 149
Hydration   10
Hydrogen   21, 42, 53, 125, 131, 132, 145, 150
Hydrogen ions   42
Hydrolysis   10, 118

I

Inorganic   108
   Acids   9, 10, 12, 151
   Compounds   156
   Nitrogen   98, 112, 113, 115
Insect
   Activity   71
   Attack   100, 130
   Control   78
   Problems   112
Insecticides   67
Insects   66, 72, 75, 78, 86
Interpreting soil test results   142
Intervention   77
Ions   147
Iron   21, 87, 126, 148
Iron potassium silicate   126
Irrigation   26, 93

K

K-Mag   127
Kelp meal   128, 130
Keratin   118

L

Labeling laws   132
Lakes   56

Leaching   17, 98
Leather meal   116, 118
Lichen   10
Liebig   22, 34
   Law of limiting factors   22
Light   77
Lignins   37, 38
Lime   46, 99, 108, 122, 124, 131, 134, 151
Limestone   131
Liquid feeding programs   120
Living mulches   96
Low-till   91
Luxury consumption   72

M

Magnesium   15, 21, 42, 43, 85, 86, 126, 127, 131, 134,
      158
   Oxide   132
   Sulfate   132
Manganese   21
Manure products   116
Manures   99, 115, 128. *See also* Green manures
MAP (monoammonium phosphate)   122
Meals   117
Meat and bone meal   116, 119
Metabolism   10
Metabolites   73
Methodologies   141
Micelles   148
Micro-nutrients   71, 76, 162
Microbial activity   46
Microorganisms   34, 86, 94, 111, 112, 116, 118, 121, 122,
      126
Milliequivalents (meqs)   154, 155
Mineral
   Nitrogen   87. *See also* Inorganic nitrogen
   Nutrients   74, 86, 145
   Salts   127
   Soils   143

Mineralization   34
Minerals   21, 73
Mining   28
Mites   67
Moisture cycle   59
Molybdenum   21
Mono-crop operations   75
Mt. Mansfield   99
Mucilages   44
Muriate of potash   125, 127
Mycorrhizae   10, 20, 68

N

Natural   107
  Defense compounds   75
  Diversity   66
  Fertilizers   119
  Inorganic   108
  Inorganic (mineral) nitrogen   117
  Minerals   108
  Nitrate of soda   117, 120
  Organic   108
  Organic fertilizers   107, 115
  Organic nitrogen   98, 113, 115
  Organic sources of nitrogen   115
  Potash   108
  Selection   66, 75
  Sources of potassium   125
  Toxins   73, 77
Negative energy   55
Nitrates   71, 72, 120, 129, 147, 151, 156, 162
Nitrification   145
Nitrites   129
Nitrogen   12, 21, 42, 46, 71, 72, 85, 87, 96, 97, 98, 100,
    107, 110, 111, 113, 114, 115, 117, 129, 133, 145, 162
  Availability   118
  Cycle   117. *See also* Cycle of nitrogen
  Efficiency   112, 115
  Fertilizer   97

Fixing bacteria   112, 146
No-till   91
Nutrient imbalances   86, 127
Nutrients   71, 74, 76, 77, 78, 86, 113, 129

O

Oak leaves   99
Oceans   56
Organic   107
   Acids   9, 10, 11, 152
   Amendments   98
   Carbon   107, 112, 115, 119, 129
   Certification   109
   Chelates   128
   Compounds   111
   Fertilizers   115
   Matter   16, 44, 45, 61, 69, 70, 72, 74, 79, 85, 86, 87,
      89, 90, 91, 93, 98, 99, 101, 110, 112, 115, 116, 120,
      128, 129, 130, 131, 142, 160
   Mulches   96
   Nitrogen   115, 118. *See also* Natural: Organic nitrogen
   Phosphorus   121
   Recommendations   142
   Residues   97
Organically grown   107
Organisms   87
Osmotic suction   60
Overdosing   72
Oxidation   145
Oxides   132, 152
Oxygen   15, 21, 42, 53, 90, 91, 96, 98, 145
Oyster shell   131
Ozone depletion   27

P

Parasites   78, 79
Parasitism   70. *See also* Parasites
Parent material   86

Particle size   97
Pathogenic fungi   69, 146
Pathogens   68, 72, 75, 90, 124
Peanut meal   119, 121
Pedosphere   18
Penicillin   73
Pest control properties   75
Pest management   67
Pesticides   25, 67, 69, 74, 76, 80
Pests   65
pH   76, 79, 90, 99, 100, 117, 118, 122, 124, 127, 131, 132,
    142, 143, 151
Phosphate   86, 92, 121, 122, 123, 124, 125, 147, 156, 157
  Ions   151
  Rock   108, 121, 122. *See also* Rock phosphate
Phosphoric
  Acid   134, 151
Phosphorus   21, 85, 119, 120, 121, 126, 143, 147, 157
Photosynthesis   77, 86
Photosynthetic organisms   18
PHSMP. *See* Buffer pH
Pine needles   99
Plant disease   66
Plant residues   111
Plant resistance   73
Plants   18, 58, 69, 72, 73, 75, 76, 77, 78, 79, 87, 93, 97,
    110, 111, 117, 157
Plows   91
Plum curculio   70
Polarity   54, 156
Pollution   122, 125
Population   139
Porosity   20, 85, 87, 92, 146
Potash   86, 125, 126, 127, 158. *See also* Potassium
Potash salts   108. *See also* Potassium; Potassium salts
Potassium   15, 21, 42, 43, 71, 73, 85, 125, 126, 143, 147,
    148, 150, 155, 157
  Chloride   125, 127
  Salts   125

Sulfate   126, 152
Potential acidity   134
Precipitated bone meal   124
Precipitated milk Phosphate   124
Precipitation   56, 72
Predator suppression   67
Predators   67, 68, 70, 71, 77, 78
Producers   18
Prometheus   24
Propagation   79
Protein   20, 71, 85, 87, 96, 98, 110, 111, 115, 117, 118, 119, 120, 121, 123, 129
Protein synthesis   96
Pruning   79
Public perception   108
Putrefaction   119

R

Rain   61, 93, 145
Rain and wind (heavy)   79
Rain forests   23
Raw bone meal   123
Recycling of nutrients   91
Reduction   156
Relative humidity   59, 60
Residues   91. *See also* Animal residues; Plant residues
Resistance   72, 73
Resistant varieties   75
Rhizosphere   20, 152
Rivers   56
Rock
   Dusts   125
   Phosphate   108, 124, 128
Root growth   86, 98
Root system   18, 19, 74, 77
Roots   8, 69, 91, 145, 150
Rotating fields   65
Rototillers   91
Rust   140

S

Sand   44, 77, 86, 158
Sandy soils   152
Saprophytes   40
Saprophytic fungi   68
Saturation   93
Seaweed   128
Seaweed extract   130
Seed
  Developement   79
  Stock   26
Semi-composting   116
Shading the soil   96
Shoot system   18
Shredding   97
Silicon   15, 147, 148
Silt   86, 158
Soap industry   123
Soap phosphate   123
Sodium   15
Sodium nitrate   117
Soft Rock Phosphate   123
Soil
  Atmosphere imbalance   127
  Colloids   57, 148
  Compaction   112
  Conditioning   113
  Conditions   77
  Conservation   75
  Depletion   79
  Fertility   71
  Horizon   58
  Organisms   71
  Respiration   67
  Sampling tubes   140
  Test laboratories   141
  Tests   99, 139
Soluble potash   133
Soluble salts   60

Solvent   53
Soybean meal   119
Space   77
Spores   68
Starches   20, 37
Steamed bone meal   123
Stomata   19, 21, 60
Straw   96
Stress   67, 71, 72, 78, 79, 98, 130
Sub-visible world   75
Suction   55
Sugars   20, 37, 71
Sul-po-mag   127
Sulfate of potash   125, 126
Sulfate of potash, magnesia   125, 127
Sulfates   147, 151, 153, 156
Sulfur   21, 85, 86, 124, 126, 127, 134, 147, 153
Sulfuric acid   124
Sun   18, 96, 110, 111, 123
Sunlight   78
Super phosphate   124
Surface area   13, 91
Survival   79
Symbiotic relationships   75
Synthetic
  Chelates   127
  Controls   112
  Inorganic   108
  Organic   108
  Organic nitrogen   115
  Pesticides   107

T

Tectonic event   17
Tectonic movement   9
Temperature   53, 72, 90, 93, 94, 96, 100
Texture   58, 62
  Analysis   158
    Triangle   158

Thatch buildup   112
Topography   62
Topsoil   58
Toxin producing organisms   74
Toxins   73, 74
Trace elements   126, 127, 128, 134
Trace minerals   122, 126
Transplanting   79
Transport systems   28
Trespass   9
Tri-calcium phosphate   122
Triple super phosphate   121, 125
Tropical soils   94
Turf   74, 77, 96
Turf diseases   114
Turning compost   90

U

Ulmins   41
Urea   36, 107, 115, 121

V

Valence   154
Vegetable protein meals   117, 119
Volatile materials   99
Volatilization   98

W

Wallerius, J.G.   34
Waste components of the feed or food industry   119
Waste stream   24
Water   10, 36, 53, 71, 77, 87, 90, 92, 93, 100, 131, 152
   Film   57
   Holding capacity   156
   Pollution   23
   Vapor   60
Waxes   37
Weather   72

Weathering   7
Weeds   66, 76, 78
   Control   77
   Cycles   78
   Seeds   90
Whey   120
Wilting point   56
Wind   8, 145
Wind-rows   91, 94
   Turners   90

Z

Zeus   24
Zinc   21, 148

"The most important fact about Spaceship Earth:
an instruction book didn't
come with it."

Buckminster Fuller

THE EDAPHIC PRESS
P.O. Box 107
Newbury, Vermont 05051

ORDER FORM

Please send me _____ copies of **EDAPHOS: Dynamics of a Natural Soil System** by Paul D. Sachs @ $14.95 each. I understand that I may return the book(s) for a full refund if I'm not satisfied.

Enclosed for book(s) is _____

VT residents, please 4% sales tax _____

Shipping (book rate) $1.50/book _____

TOTAL ENCLOSED _____

Name _____

Business Name _____

Address _____

City _____ State _____

Zip _____ Phone _____